VEGETABLE TROUBLES

This book reveals the vast array of troubles which can affect the vegetable plot. Some of them are pests and diseases, but not all — split tomatoes, blown sprouts and bolted onions do not appear in the pest charts but they are nevertheless important vegetable troubles. It is not the intention of this book to frighten you; no matter how long you garden you will never see all the troubles. On the contrary the purpose of this guide is to take away the worry of the unidentified problem, and to provide you with the knowledge to deal with the trouble speedily and correctly.

There is perhaps nothing more distressing in gardening than to see a whole crop wiped out – eaten by a pest rather than by you. There is no vet you can take your sick plant to, so you must learn to be your own vegetable doctor.

Types of vegetable troubles

PESTS
A pest is a member of the animal kingdom which attacks plants. Nearly all are insects (small creatures with 6 legs at the adult stage) and here are found the caterpillars, beetles and flies. A few pests, such as mites, woodlice and millepedes are not really insects, but are often included in the 'insect' group. Some pests (e.g. the microscopic eelworm) are much smaller than insects. Other pests, such as cats and birds, are much larger.

DISEASES
A disease is a vegetable trouble caused by a living organism which is transmitted from one plant to another. For disease to occur the minute organisms must be present and the condition must favour infection. Good cultural practice is designed to avoid these conditions wherever possible. Most diseases are caused by fungi, and these can often be prevented by spraying. The others, caused by bacteria and viruses, can rarely be controlled in this way.

DISORDERS
A disorder is a vegetable trouble which may have disease-like symptoms, but it is not caused by a living organism. Disorders indicate that something is or has been wrong with the growing environment. This may not be a matter of 'good' or 'bad' conditions – cool air and spongy soil are ideal for certain crops but produce serious disorders in others. There are some conditions which are generally damaging and these are listed on page 4.

Golden Rules for staying out of trouble

● **LEARN HOW TO GROW VEGETABLES**
Don't rely on just reading the seed packet. Buy a book – there are lots to choose from. The Vegetable Plotter provides essential information in charts and pictures.

● **ROTATE YOUR CROPS**
Soil troubles and deficiencies will build up if you grow the same crop year after year on the same plot of land. Crop rotation is vital for continued plant health.

● **PREPARE THE SOIL PROPERLY**
Good drainage is vital. The time for digging is autumn or early winter if you plan to sow in spring. Follow the rules in your vegetable book for the correct way to manure, feed and lime the soil.

● **SOW THE RIGHT SEED**
Always buy good quality seed and choose carefully. Don't just choose, 'cabbage' or 'cauliflower' – make sure the variety is suitable for the chosen sowing date.

● **CHOOSE GOOD TRANSPLANTS**
A poor seedling will never develop into a good plant. When buying transplants make sure they are sturdy, green and with a good root system.

● **AVOID OVERCROWDING**
Sow seed thinly. Thin the seedlings as soon after germination as possible. Overcrowding leads to crippled plants and high disease risk.

● **GET RID OF WEEDS & RUBBISH**
Weeds rob the plants of water, food, space and light. Rubbish, like weeds, can be a breeding ground for pests and diseases.

● **GET RID OF BADLY INFECTED PLANTS**
Do not leave sources of infection in the garden. Remove and destroy incurable plants when this book tells you to do so.

● **FEED & WATER CORRECTLY**
Many plant troubles are due to incorrect feeding and soil moisture problems. Use a balanced fertilizer in order to avoid soft growth. Never let the roots get dry but daily sprinklings instead of a good soaking will do more harm than good.

● **USE A SPRAY WHEN NECESSARY**
There are times when a spray or some other chemical treatment may be necessary – this book will tell you which one to use. Inspect the plants regularly and keep a general purpose insecticide, such as Crop Saver, handy. There are a few simple rules to ensure effective, safe and economical pest control: Before spraying read the instructions – avoid using too little or too much. Apply a fine drenching spray above and below the leaves. After spraying wash out the sprayer and do not keep the liquid for use next week.
Insecticides are normally applied at the first sign of attack. Systemic products go inside the sap stream and protect parts not reached by the spray. Fungicides are usually protectants, and need to be applied **before** trouble appears.
Some troubles (red spider mite, whitefly, diseases etc.) need repeated spraying. Once again, follow the instructions on the packet. Finally, choose a product with a suitable harvest interval: during the picking season choose a product with a 0–2 day interval between spraying and harvesting.

Causes of General Disorders

Some vegetable troubles attack a single or small group of crops — potato blight, carrot fly and club root are examples. Others are more general, capable of attacking most vegetables. The important general troubles are described on these two pages.

WIND

Wind is often ignored as a danger, yet a cold east wind in spring can kill in the same way as frost. More frequently the effect is the browning of leaf margins. Another damaging effect is wind rock, which can lead to rotting of the roots.

FROST

A severe late frost will kill half-hardy vegetables. The shoots of asparagus and potatoes are blackened, but healthy shoots appear after the frosts have passed. The general symptoms of moderate damage are yellow patches or marginal browning of the leaves. The basic rule is to avoid sowing or planting before the recommended time unless you can provide protection. If your garden is on a sloping site, open part of the lower boundary to air movement or you will create a 'frost pocket'.

TOO LITTLE WATER

The first sign is a dull leaf colour, and this is followed by wilting of the foliage. Discoloration becomes more pronounced and growth is checked. Lettuces become leathery, roots turn woody and some plants run to seed. Flowers and young fruit may drop off. If water shortage continues, leaves turn brown and fall, and the plant dies. Avoid trouble by incorporating organic matter, by watering thoroughly and by mulching.

TOO MUCH WATER

Waterlogging affects the plant in two ways. Root development is crippled by the shortage of air in the soil. The root system becomes shallow, and also ineffective as the root hairs die. Leaves often turn pale and growth is stunted. The second serious effect is the stimulation of root-rotting diseases. Good drainage is therefore essential, and this calls for thorough autumn digging. Incorporate plenty of organic matter into heavy soil.

HEAVY RAIN FOLLOWING DROUGHT

The outer skin of many vegetables hardens under drought conditions, and when heavy rain or watering takes place the sudden increase in growth stretches and then splits the skin. This results in the splitting of tomatoes, potatoes and roots. This can be avoided by watering before the soil dries out.

TOO LITTLE PLANT FOOD

The major plant foods are nitrogen, phosphate and potash, and a vigorous crop acts as a heavy drain on the soil's resources. Nitrogen shortage leads to stunted growth, pale leaves and occasional red discoloration. Potash shortage leads to poor disease resistance, marginal leaf scorch, and produce with poor cooking and keeping qualities. Before sowing or planting apply a complete fertilizer, such as Growmore or Crop Booster, containing all the major nutrients.

With all crops one or more dressings should be applied to the growing plants. Backward vegetables are helped by a leaf-feeding fertilizer such as Fillip.

SHADE

In a small garden deep shade may be the major problem. Straggling soft growth is produced and the leaves tend to be small. Such plants are prone to attack by pests and diseases. Grow leaf and root types rather than fruit and pod vegetables.

TRACE ELEMENT SHORTAGE

Vegetables often show deficiency symptoms such as yellowing between the veins and leaf scorch. The most important trace elements are magnesium, manganese, iron, boron and molybdenum. Make sure the soil is well supplied with compost or manure. Foliar feeding will help, and so will sequestered compounds. If you are sure that an element is deficient, you can apply a curative treatment:

Magnesium: Spray with Epsom salts (3 oz per gallon of water) every 2 weeks.

Boron: Borax (1 oz raked into 20 sq. yards).

Manganese: Manganese sulphate ($\frac{1}{4}$ oz per gallon of water per 10 sq. yards).

TOO LITTLE ORGANIC MATTER

The soil must be in good heart and this calls for liberal amounts of organic matter. Not all materials are suitable; peat may increase aeration and water retention but the need is for an active source of humus. Good garden compost and well-rotted manure are ideal. Timing is all-important — at least a season before root crops or brassicas, at digging time for other crops.

General Pests and Diseases

APHID

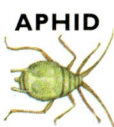

The weakening effect of greenfly and blackfly on leaves and shoots is obvious. There are, however, other damaging results. Sticky honeydew is deposited, and the sooty moulds which grow on it are unsightly and block the leaf pores. Even worse is the danger of virus infection, as aphids are the prime carriers. For these reasons aphids should be tackled quickly. Outdoors spray with Topgard Systemic Liquid or Crop Saver; under glass use Crop Saver.

EARWIG

The leaves of beetroots, parsnips and carrots may be skeletonized by this pest. Spray with Crop Saver when they are first noticed.

BIRDS

Birds are a joy in the garden and most of them do no harm. A few species, however, are a serious nuisance to seeds, seedlings and some mature crops, and netting is necessary.

CATS

Cats often choose seed beds for toilet purposes, and usually avoid their own gardens. This is a difficult problem and Pepper Dust should be sprinkled liberally where cats are a nuisance.

SOIL PESTS
GROUP 1: Controlled by Slug Gard

Slug Gard, based on methiocarb, is an alternative to metaldehyde for the control of slugs. It has the advantage of being effective in wet as well as dry weather. Research has shown that a light sprinkling raked into the soil surface will control woodlice, millepedes and leatherjackets.

SLUGS & SNAILS

Extremely troublesome pests especially in wet weather. Seedlings may be killed; leaves, stems and roots of older plants are damaged. Look for the tell-tale slime trails.

LEATHERJACKET

Dark grey grubs, about 1 in. long. Most active in light soils and wet weather. Stems are attacked, lower leaves devoured. Root crops are tunnelled.

MILLEPEDE

Pink or black grubs which curl up when disturbed. They attack underground parts of plants, often extending areas damaged by other pests. Most troublesome under cool, damp conditions. Not easy to control.

WOODLICE

Grey, hard-coated pests found in greenhouses and frames. Seedlings and young plants are attacked. They hide under pots, rubbish etc. during the day.

SOIL PESTS
GROUP 2: Controlled by Bromophos

Bromophos can be used for all vegetables without risk of taint. Apply $\frac{1}{2}$ oz per sq. yard and lightly rake in.

CUTWORM

Fat grey or brown caterpillars, $1\frac{1}{2} - 2$ in. long. They live near the surface and eat young plants at ground level. Stems are often severed.

CHAFER GRUB

Large curved grubs, over 1 in. long. They feed throughout the year on roots and are a pest in newly broken-up grassland.

WIREWORM

Shiny, slow-moving grubs. They attack a wide variety of vegetables. Stems are gnawed below ground, root crops are tunnelled.

DAMPING OFF

Germinating seedlings can be attacked by the damping off fungi, withering and blackening at the base before toppling over. Indoors use sterilized compost, sow thinly, water carefully, ventilate properly and provide adequate light. Outdoors avoid sowing in cold wet soil, sow thinly and do not overwater. If the disease does occur remove the affected seedlings immediately and water remainder with Cheshunt Compound.

PESTICIDE	Recommended Interval between Spraying and Harvesting
Benlate	0 days
Thiophanate-methyl	0 days
Bio Sprayday	0 days
Malathion	1 day
Derris	1 day
Crop Saver	2 days
Topgard Systemic	7 days
Bromophos	7 days
Slug Gard	7 days
Pirimiphos-methyl	7 days
Dimethoate	7 days
Carbaryl	7 days
Dithane	7–14 days
Diazinon	14 days
Fenitrothion	14 days
Hexyl Plus	14 days
Pirimicarb	14 days
Lindane	14 days

Beans & Peas

WHAT'S WRONG...

The major threat to the crop depends upon the type of bean you grow. Black bean aphid is the main danger to broad beans, with chocolate spot as the most serious disease. The chief disorder of runner beans is the failure of pods to set, and the pea crop is threatened by birds and the pea moth. French beans can be attacked by numerous pests and diseases, but these are rarely serious.

	Symptom	Likely causes
Seeds & Seedlings	— missing	3 or 16
	— little or no germination	1 or 2 or 8 or **Millepede** or **Damping off** (see page 5)
	— tunnelled before sowing	1
	— tunnelled after sowing	2
Stems	— brown streaks outside	13 or 20
	— brown streaks inside	18
	— brown or purple spots	12 or 21
	— wilted, dying	8 or 18
	— mouldy	17
	— infested with aphids	4 or 5
	— brown or blackened at base	8
Leaves	— yellowed or yellow patches	6 or 8 or 15 or 18
	— silvery patches	10
	— spotted	12 or 13 or 20 or 21
	— notched	9
	— holed	**Slugs & Snails** (see page 5)
	— white, mauve or brown mould	6 or 7
	— infested with aphids	4 or 5
	— bronzed, speckled	**Red spider mite** (see page 19)
Flowers	— absent	11
	— present, but pods absent	14
	— damaged	4 or 5 or 14 or 16
Pods	— distorted	4 or 6 or 10
	— torn	16
	— spotted, dry texture	6 or 13 or 20 or 21
	— spotted, wet texture	12
	— spotted, mouldy texture	7 or 17
	— silvery patches	10
Beans & Peas	— tunnelled, maggots present	19
	— brown spot in centre	15

TUNNELLED SEEDS

1 SEED BEETLE

Seeds of peas and beans are sometimes found to bear small, round holes. Within these tunnels are the tiny seed beetle grubs. Affected seeds do not germinate or produce weak seedlings.

Treatment: None.

Prevention: Buy good quality seed. Never sow seeds if they are holed.

TUNNELLED SEEDLINGS

¼ in. white grubs

2 BEAN SEED FLY

All bean varieties are susceptible to attack by these soil-living grubs. Damaged seeds fail to germinate; tunnelled seedlings wilt and become distorted. Early crops are worst affected.

Treatment: Destroy damaged seedlings.

Prevention: Prepare a good seed bed. Dust the seed drills with Bromophos.

3 MICE

Mice can be serious pests, as they are capable of clearing whole rows of pea seeds and seedlings overnight. Old fashioned remedies are to dip the seed in paraffin or alum, or to put spiny branches along the rows. A mouse bait, such as Racumin, can be used if the site is known to have a mouse problem.

4 BLACK BEAN APHID

A serious pest of broad beans in spring and french beans in July and August. Large blackfly colonies stunt growth, damage flowers and distort pods.

Treatment: Spray with Crop Saver or Topgard Systemic Liquid at the first signs of attack. Repeat as necessary.

Prevention: Pinch out the tops of broad beans once four trusses of pods have formed.

APHIDS ON LEAVES

5 PEA APHID

Not often a serious pest, but in a hot, damp summer large colonies can severely damage peas. Growth is stunted and flowers are damaged.

Treatment: Spray with Crop Saver or Topgard Systemic Liquid at the first signs of attack.

Prevention: No practical method available.

6 DOWNY MILDEW

Yellowish blotches on the leaves of peas, with a pale mauve or brown mould on the underside. Attacks occur in cool, wet seasons. Infected pods are spotted and distorted.

Treatment: Spray with Dithane at the first signs of disease. Repeat at fortnightly intervals.

Prevention: Practise crop rotation. Burn affected plants after picking.

MOULDY LEAVES

7 POWDERY MILDEW

White powdery patches appear on both sides of the leaves of peas. Attacks occur in dry seasons and are worst in sheltered gardens. Infected pods are covered with white patches.

Treatment: Spray with Benlate at the first signs of disease. Repeat at fortnightly intervals.

Prevention: Burn affected plants after picking.

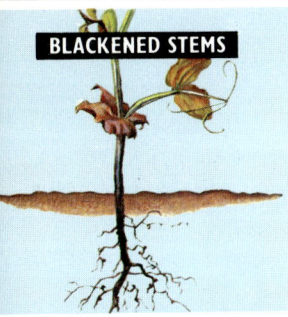

BLACKENED STEMS

8 FOOT ROT & ROOT ROT

Leaves turn yellow and shrivel; roots and stem bases turn brown or black and soon start to rot.

Treatment: Lift and burn badly affected plants. Water soil with Cheshunt Compound to check the spread of the disease to other plants.

Prevention: Rotate crops.

NOTCHED LEAVES

¼ in. brown beetles

9 PEA & BEAN WEEVIL

Tell-tale signs are U-shaped notches at the edges of the young leaves. Growth is retarded but older plants generally soon recover. Seedlings, however, can be killed by a severe attack.

Treatment: Spray with Fenitrothion at the first signs of attack.

Prevention: Hoe around the plants in April and May.

SILVERY PODS

10 PEA THRIPS

Silvery patches appear on leaves and pods. The pods are distorted and the yield is reduced. Attacks are worst in hot, dry weather. Minute black or yellow insects are just visible.

Treatment: Spray with Fenitrothion or Crop Saver.

Prevention: Dig over the soil after removing an infected crop.

11 NO FLOWERS

Peas and beans sometimes fail to produce flowers. This uncommon complaint can be due to a severe infestation of capsid bug (see page 5) or pea thrips which causes the flower buds to wither. But the most likely cause of a shortage of flowers is the presence of too much nitrogen in the soil. Always use a balanced fertilizer containing phosphates and potash for peas and beans.

Golden Rules for staying out of trouble

Buy good quality seed. Enrich the soil with compost. Mulch the growing plants. Water thoroughly in dry weather. Pick pods regularly. Choose a fresh site next year.

12 HALO BLIGHT

Small brown spots on leaves, each one being surrounded by a yellow 'halo'. Pods develop water-soaked spots. Plants are stunted and yields are reduced. Attacks are worst in a wet season.

Treatment: Lift and destroy diseased plants.

Prevention: Practise crop rotation. Never soak seed before sowing.

SPOTS ON LEAVES

French beans, Runner beans

Broad beans

13 CHOCOLATE SPOT

Small brown spots on leaves; dark streaks along the stems. Pods may be affected and the seeds discoloured. In a bad attack the spots join together and the plant is killed.

Treatment: Lift and destroy diseased plants. Spray remaining plants with Benlate.

Prevention: Apply a vegetable fertilizer before sowing and do not grow the plants too closely together.

Beans & Peas
continued

BROWN-CENTRED PEAS

14 NO PODS

One of the major problems with runner beans is their tendency to lose their flowers without forming pods. Sparrows can be the culprits, and so can bumble bees. Cool weather at flowering time results in a lack of pollinating insects, but the failure of beans to set is always worst in a hot, dry season. Keeping the roots moist by digging in compost, by mulching and watering is helpful, but recent research has shown that the practice of spraying the flowers is of little value. The best way to avoid trouble is to grow a white- or pink-flowering variety.

15 MARSH SPOT

Tell-tale sign is a brown-lined cavity in the centre of each pea. The cause is a shortage of manganese in the soil. The only outward sign is a slight yellowing between the leaf veins.

Treatment: None.

Prevention: Incorporate compost into the soil before sowing. Apply a sequestered compound. Repeated spraying with Fillip may help.

16 BIRDS

Birds can be a nuisance in several ways. Pigeons devour pea seeds and seedlings, and sparrows will tear open pea pods. Sparrows will also damage runner bean flowers and so prevent pods from being produced. Scarers and cotton are not really effective; netting is by far the best answer.

ROTTEN PODS

17 GREY MOULD (Botrytis)

The pods of french beans and occasionally peas may develop a grey velvety mould in wet weather. This mould may also coat the stem surface.

Treatment: Pick and burn affected pods. Spray with Benlate to protect remaining pods.

Prevention: Spray with Benlate at flowering time if grey mould is a recurrent problem in your garden.

BROWN-STREAKED TISSUE

18 FUSARIUM WILT

Outward signs are stunted growth, yellowing or rolled leaves and little or no crop. If you cut open the stem of an infected plant the tell-tale signs of wilt are revealed. Reddish-brown longitudinal streaks run through the stem tissue, but no external browning occurs.

Treatment: Remove and burn affected plants.

Prevention: Grow wilt-resistant varieties.

MAGGOTY PEAS

⅓ in. greenish maggots

19 PEA MOTH

Maggoty peas are well known to all vegetable growers, especially in S. England. Pea moth maggots burrow through the pods and into the seeds, making them unusable. Early and late sown crops often escape damage.

Treatment: None.

Prevention: Spray with Fenitrothion 7-10 days after the start of flowering.

20 ANTHRACNOSE

Brown sunken spots on pods. Stem cankers appear and leaves bear brown patches. At a later stage these brown spots and patches may turn pink. In a bad attack the plant may be killed.

Treatment: Lift and destroy diseased plants. Spray remaining plants with Benlate.

Prevention: Practise crop rotation. Dust seed with Benlate before sowing.

DRY SPOTS ON PODS

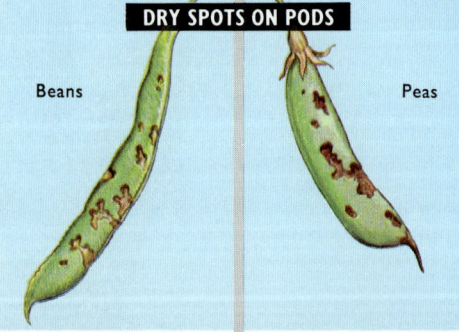

Beans

Peas

21 LEAF & POD SPOT

Brown sunken spots on pods. Peas may be discoloured. Leaves and stems bear similar brown spots and in a bad attack they may join together. Early crops suffer most, especially in a wet season.

Treatment: Lift and destroy diseased plants.

Prevention: Practise crop rotation. Dust seed with Benlate before sowing.

Beetroot

WHAT'S WRONG...

Beetroot is an easy crop to grow, and is generally trouble-free. Black bean aphid and mangold fly are occasional nuisances, but yields are not usually seriously affected. You may find that the leaves are discoloured; beetroot is one of the most sensitive indicators of trace element deficiency in the soil.

	Symptom	Likely Causes
Seedlings	— eaten	**Birds** or **Slugs** (see page 5)
	— toppled over	**Damping off** (see page 5)
	— blackened	**3**
Leaves	— blistered	**1**
	— rolled	**5**
	— spotted	**6**
	— infested with blackfly	**Black bean aphid** (see page 6)
	— mouldy patches	**Downy mildew** (see page 27)
	— mottled	**5**
Plants	— run to seed	**4**
Roots	— small and leathery	**Dry soil** or **Fertilizer shortage**
	— large and leathery	**Delayed harvesting**
	— blackened inside, cankered outside	**2**
	— eaten	**Swift moth** (see page 15) or **Cutworm** or **Millepede** (see page 5)
	— covered with purple mould	**Violet root rot** (see page 15)
	— scabby patches	**Common scab** (see page 26)
	— split	**Splitting** (see page 15)

BLISTERED LEAVES

1 MANGOLD FLY (Leaf Miner)

Small white grubs burrow inside the leaves, causing tunnels which later turn into blisters. Attacks occur from May onwards, and the effects are most serious on young plants. Badly damaged leaves turn brown and growth is retarded.

Treatment: Pick off and destroy affected leaves. Spray with Malathion or Crop Saver at the first sign of attack.

Prevention: None.

BLACKENED ROOTS

2 HEART ROT

Leaves wilt in the summer and the tops of the roots develop brown, sunken patches. A cut root reveals blackened areas within the flesh. The cause is boron deficiency, and attacks are worst on light, over-limed land in a dry season.

Treatment: Repeated spraying with Fillip may help.

Prevention: If soil is known to be boron deficient, apply 1 oz borax per 20 sq. yards before planting – take care not to overdose.

3 BLACK LEG

Black leg is a serious disease of seedlings, causing them to turn black and shrivel. It occurs when seeds are sown too thickly in compacted soil which becomes waterlogged in wet weather. If an attack occurs, remove diseased plants and water remainder with Cheshunt Compound.

4 BOLTING

Plants sometimes run to seed before roots have developed. Dry soil or a shortage of organic matter is the usual cause, but it will occur if you sow too early or if you wait too long before thinning the seedlings. Grow a resistant variety, such as Boltardy, if bolting has been a problem.

ROLLED LEAVES

5 SPECKLED YELLOWS

Yellow patches develop between the veins, and in a severe attack the whole leaf turns yellow and then brown. The tell-tale sign is the inward rolling of the leaf edges. The cause is manganese deficiency.

Treatment: Apply a sequestered compound. Repeated spraying with Fillip may help.

Prevention: Do not overlime the soil.

SPOTTED LEAVES

6 LEAF SPOT

Brown spots appear on the leaves. The pale central area of each spot sometimes drops out. The foliage may be badly disfigured by these numerous small spots but the effect on yield is not serious.

Treatment: None. Pick off and destroy badly diseased leaves.

Prevention: Practise crop rotation. Apply a balanced fertilizer, such as Crop Booster, before sowing seed.

Golden Rules for staying out of trouble

Leave a season between manuring and growing beetroot. Avoid over-liming. Prepare a good seed bed. Sow thinly and thin out seedlings early. Take care when hoeing. Water in dry weather.

The Brassica Family

CABBAGE · CAULIFLOWER · KALE
BRUSSELS SPROUTS · BROCCOLI · KOHL RABI

Kale is generally a trouble-free vegetable, but the other popular members of the brassica family are subject to a wide range of pests, diseases and other disorders. The worst troubles are cabbage root fly, cabbage caterpillars, mealy aphid, whitefly, club root, flea beetle and pigeons.

WHAT'S WRONG...

	Symptom	Likely Causes
Seedlings	– eaten	**11** or **13** or **16** or **19**
	– toppled over	**6** or **Damping off** (see page 5)
	– peppered with small holes	**16**
	– severed at ground level	**19**
Stems	– tunnelled, maggots present	**28**
	– blackened zone near soil level	**6**
Leaves	– swollen, distorted	**18**
	– narrow, strap-like	**9**
	– curled, blistered	**20**
	– whitened	**17**
	– diseased	**1** or **5** or **7** or **8**
	– coloured between green veins	**22** or **23**
	– holed	**10** or **11** or **13** or **16** or **26**
	– infested with greenfly	**20**
	– tiny white moths	**21**
	– caterpillars	**10** or **26**
Roots	– swollen	**3** or **4**
	– tunnelled, maggots present	**2**
	– eaten	**24**
Plants	– bluish leaves, wilting in sunshine	**2** or **4**
	– blind; plants not growing	**9** or **18** or **Blind transplants**
	– wilting, dying	**24** or **28**
Brussels sprouts	– buttons open and leafy	**15**
Cabbage	– no hearts	**12**
	– split hearts	**14**
Cauliflower	– small heads	**9** or **27**
	– brown heads	**25**

1 DOWNY MILDEW

Yellowing of upper surface. Wh furry fungus growth beneath. Usua restricted to young plants; ove crowding and a moist atmosphere e courage its spread. Growth severe checked.

Treatment: Spray with Dithane the first sign of disease.

Prevention: Sow seeds in steriliz compost. Choose a fresh site raising seedlings if downy mildew been a problem in the past.

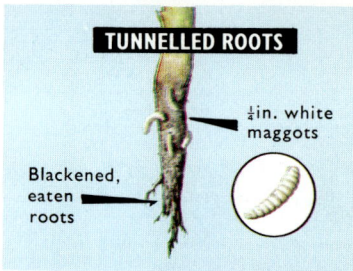

TUNNELLED ROOTS

½in. white maggots

Blackened, eaten roots

2 CABBAGE ROOT FLY

Tell-tale signs are blue-tinged leaves which wilt in sunny weather; recent transplants are particularly susceptible. Young plants die; older ones grow slowly and develop badly. Cabbages fail to heart, cauliflowers form tiny heads.

Treatment: A thorough spray around the base of plants with Crop Saver may give partial control.

Prevention: Sprinkle Bromophos around the base of plants immediately after transplanting if this pest has been a problem.

3 GALL WEEVIL

Much less serious and much less common than club root. Swellings generally form close to ground level. Growth may be checked slightly, but there is rarely any serious effect on yield.

Treatment: Not worth while. Spread can be reduced by watering around plants with Crop Saver.

Prevention: Soil-pest killers which have been used to prevent more serious problems may give some control of gall weevil.

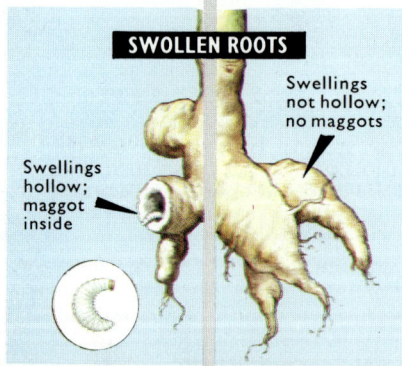

SWOLLEN ROOTS

Swellings not hollow; no maggots

Swellings hollow; maggot inside

4 CLUB ROOT (Finger and Toe)

Tell-tale signs are discoloured leaves which wilt in sunny weather. A serious disease which can be disastrous in a wet season. Plants may die or grow very slowly.

Treatment: None. Lift diseased plants and burn. In case of a severe attack do not plant brassicas on the site for several years.

Prevention: Make sure the land is adequately limed and well drained. Dip transplant roots in Benlate solution or Calomel Dust paste before planting.

WHITE BLISTER (White Rust)

...hite spots appear on leaves. The fungal ...sses may spread in a mild, damp season ...form a white felt over the leaf surface. ...owth is stunted and plants may die.

...eatment: Cut off and burn diseased ...ves. Thin out plants to reduce over-...owding.

...evention: Do not grow brassicas on ...d affected in the previous season.

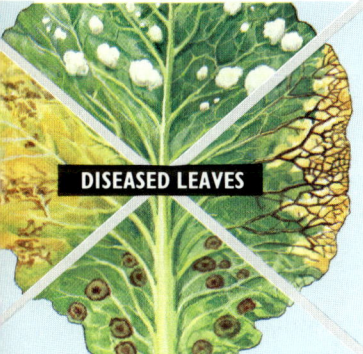

DISEASED LEAVES

6   WIRE STEM

Base of stem becomes black and shrunken. Seedlings often die; survivors grow very slowly and stems break easily.

Treatment: None.

Prevention: Avoid growing seedlings in wet and cold soil or compost. Cheshunt Compound may help. Avoid over-crowding.

SHRUNKEN STEMS

7   BLACK ROT

An uncommon but serious disease. Seedlings are killed; mature plants are severely stunted, bearing yel-low leaves and characteristically black veins. Lower leaves generally fall. If the stem is cut across a dark brown ring is revealed. Worst attacks occur in a warm, wet summer.

Treatment: None. Lift diseased plants and burn.

Prevention: None. Rotate crops.

Golden Rules for staying out of trouble

Leave several months between digging and planting. Lime if necessary. Rake in Bromophos before planting. Remember to plant firmly. Spray as soon as pests are seen.

LEAF SPOT (Ring Spot)

...wn rings up to 1 inch across appear on ...ture leaves. Badly infected foliage may ...n yellow and fall. Most likely to occur ...S.W. areas. It is encouraged by wet ...ather.

...eatment: Cut off and burn diseased ...ves. Spray with Dithane to stop it ...eading.

...evention: Do not grow brassicas on ...d affected by leaf spot in the previous ...son.

9   WHIPTAIL

Leaves are thin and strap-like. Plant growth is poor; cauliflower heads may be very small or fail to develop. Caused by molybdenum deficiency due to acid soil.

Treatment: Repeated spraying with Fillip.

Prevention: Make sure soil is adequately limed before sowing or planting.

NARROW LEAVES

10   CABBAGE CATERPILLARS

...ook for cabbage caterpillars when holes ...egin to appear in leaves. The cabbage ...oth tends to burrow into the heart. The ...sk period is April – October, and ...ttacks are worst during a hot dry ...ummer and in coastal areas.

...reatment: Spray with Crop Saver as ...oon as the first attacks occur. Repeat ...s necessary.

...revention: Inspect underside of leaves ...f you see white butterflies hovering over ...e plants. Remove and crush any eggs ...hich have been laid.

LARGE HOLES IN LEAVES

SMALL CABBAGE WHITE — velvety

LARGE CABBAGE WHITE — hairy

CABBAGE MOTH — smooth

11   PIGEONS

Pigeons are a serious pest in many areas, stripping away the soft portions of the leaves until only the stalks and veins remain. Droppings make the produce tedious to prepare for cooking. Troublesome throughout the year, especially in winter.

Treatment: None.

Prevention: Bird scarers are of limited value. Use nylon netting, making sure that the plants are completely enclosed.

12   HEARTLESS CABBAGES

...here are several reasons why ...bbage plants fail to heart. Too little ...rganic matter in the soil and too ...ttle compaction of the ground before ...anting are common reasons. So is ...iling to plant the seedlings firmly. ...rought increases the risk; so does a ...ady site. Proper feeding will help, ...ut use a balanced fertilizer (Crop ...ooster or Growmore) and not ...raight nitrogen.

13   SLUGS & SNAILS

Leaves and stems may be severely attacked during wet weather. The pests are generally not seen during the day, so look for the tell-tale slime trails. Young plants are particularly susceptible and may be killed.

Treatment: Scatter Slug Gard or Slug Pellets around the plants at the first signs of attack.

Prevention: Keep surrounding area free from rubbish.

14   SPLIT-HEADED CABBAGES

There are two major reasons why cabbage heads suddenly split. In summer the usual cause is rain after a long period of dry weather. Foliar feeding with Fillip at the first sign of trouble helps to harden the leaf tissue, but it is better to prevent trouble by watering regularly during drought. In winter the cause is a sudden sharp frost; consider lifting and storing mature heads if very cold weather is forecast.

The Brassica Family
continued

15 BLOWN BRUSSELS SPROUTS

Brussels sprouts sometimes produce open, leafy sprouts instead of hard, round buttons. These blown sprouts should be removed promptly. The causes are similar to the factors responsible for heartless cabbages – not enough well-rotted organic matter in the soil, too little consolidation of the ground before planting and failure to plant firmly. Make sure the plants are kept well watered during dry-weather and avoid planting too closely. Choose an F_1 hybrid variety.

SMALL HOLES IN LEAVES

16 FLEA BEETLE

A serious pest, especially in April and May during warm, settled weather. Young leaves bear numerous, small round holes. Growth is slowed down and seedlings may be killed. The tiny beetles jump when disturbed.

Treatment: Spray the seedlings with Crop Saver as soon as the first signs of damage are noticed. Water damaged plants if the weather is dry.

Prevention: Treating seed with an insecticidal seed dressing before sowing will prevent early attacks.

WHITENED LEAVES

BLIND PLANTS

Growth stopped

17 FROST

Frost can be extremely damaging to brassicas. If the plants are not firmly anchored in the soil then frost can reach the roots and lead to the death of the plant. Frost can also damage the leaves of non-hardy varieties. The blanched areas are quickly attacked by fungi or bacteria and extensive rotting can occur.

Treatment: Remove and burn damaged leaves.

Prevention: Always plant firmly. Frost damage is worst on soft growth so always use a properly balanced fertilizer when preparing the soil.

18 SWEDE MIDGE

Attacks are uncommon, but the results are devastating. Leaf stalks near the growing point are swollen and distorted, and the plants become blind. Look carefully for the minute white larvae on the leaf stalks. Northern and S.W. districts are most at risk.

Treatment: Lift and burn badly affected plants. Spraying with Hexyl Plus as soon as the first signs are noticed may stop it spreading to other plants.

Prevention: No practical method available.

19 CUTWORM

These large grey or brown caterpil live just below the surface. Young pla are attacked at night and stems severed at ground level. Leaves and r may also be eaten. Plants are most at in June and July.

Treatment: Hoe the soil around plants regularly during the dar months. Pick out and destroy caterpi which are brought to the surface.

Prevention: Rake Bromophos into soil before planting.

20 MEALY APHID

Large clusters of waxy, greyish 'greenflies' occur on the underside of leaves and tips of plants from June onwards in hot, dry weather. Affected leaves curl and turn yellow. Sooty moulds develop if attack is severe.

Treatment: Not easy to control. Spray thoroughly at the first sign of attack with Topgard Systemic Liquid or Crop Saver.

Prevention: Dig up and destroy old cabbage and Brussels sprouts stalks.

INSECTS ON LEAVES

21 CABBAGE WHITEFLY

Outbreaks have become much more widespread in the past few years. Tiny white moths and larvae feed on the underside of leaves. Affected plants are weakened and sooty moulds develop. The adults are active throughout the year and fly into the air when disturbed.

Treatment: Not easy to control. Spray with Crop Saver or Sprayday at 3 day intervals until the infestation has been cleared. Best results are obtained by spraying in the morning or evening.

Prevention: No practical method available.

22 MAGNESIUM DEFICIENCY

Yellowing between the leaf veins begins on the older leaves, and these yellow areas may eventually turn orange, white, red or purple. Magnesium deficiency is much more common than manganese deficiency.

Treatment: Apply a sequestered compound. Repeated spraying with Fillip may help.

Prevention: Incorporate compost into the soil during autumn digging. Use a fertilizer containing magnesium.

MOTTLING BETWEEN VEINS

23 MANGANESE DEFICIENCY

It is not always easy to distinguish between manganese and magnesium deficiency by looking at a single leaf. Manganese deficiency symptoms, however, usually start on young as well as old leaves, and the leaf edges are often incurled and scorched.

Treatment: Apply a sequestered compound. Repeated spraying with Fillip may help.

Prevention: Incorporate compost into the soil during autumn digging.

24 CHAFER GRUBS

The visual symptoms of chafer grub attack are wilting leaves and dying plants. On lifting affected specimens damaged roots are seen and fat, curved grubs may be found in the soil. These slow-moving pests can feed throughout the year. Gardens made from newly-dug grassland are the areas most likely to suffer.

Treatment: None.

Prevention: Hand pick and destroy the grubs when autumn digging. Rake Bromophos into the soil before planting.

ROOTS EATEN

1½ in. soil-living grubs

BROWN CURDS

25 BORON DEFICIENCY

Cauliflowers are extremely sensitive to boron deficiency in the soil. Young leaves are distorted, and the heads are small and bitter. The main symptom is the development of brown patches on the curds.

Treatment: Repeated spraying with Fillip may help.

Prevention: Incorporate compost into the soil during autumn digging. If soil is definitely known to be boron deficient, apply 1 oz borax per 20 sq. yards before planting – take care not to overdose.

SEVERED STEMS

1-2 in. soil-living caterpillars

SKIN-COVERED HOLES

½ in. green caterpillars

TUNNELLED STEMS

Cut stem reveals ¼ in. grubs

27 BUTTON CAULIFLOWERS

Many gardeners fail to grow satisfactory cauliflowers. Buttoning often takes place, which is the production of very small heads early in the season. These button heads quickly run to seed. Unfortunately there are many causes. It may be due to an early attack by a pest or disease, such as whiptail, flea beetle or boron deficiency. Or the cause may be cultural – poor soil, insufficient consolidation of the ground before planting, loose planting, drought or failure to harden off the seedlings properly before planting.

26 DIAMOND-BACK MOTH

These green caterpillars can be a serious summer nuisance in coastal areas. They feed on the underside of the foliage and, unlike the cabbage caterpillars, generally leave the upper skin intact. When disturbed they drop from the plant on a silken thread. In a severe attack the leaf is completely skeletonized.

Treatment: Spray with Crop Saver as soon as the first attack occurs.

Prevention: None.

28 CABBAGE STEM FLEA BEETLE

Infested plants wilt and die. Cut open the stem of one of the plants if you suspect cabbage stem flea beetle; the tell-tale sign is the presence of small, cream-coloured grubs. Attacks occur between August and October.

Treatment: Lift and burn infested plants.

Prevention: Do not grow brassicas on land affected by cabbage stem flea beetle in the previous season.

Carrots & Parsnips

WHAT'S WRONG...

	Symptom	Likely Causes
Seedlings	– failed to appear	**Sowing too deeply**
	– toppled over	**Damping off** (see page 5)
Leaves	– mottled yellow, later red	**9**
	– reddish, later yellow	**1**
	– badly distorted	**10**
	– collapsed, leaf stalks black	**7**
	– tunnelled, blistered	**Celery fly** (see page 16)
	– covered with blackfly	**Black bean aphid** (see page 6)
Plants	– toppled over, brown affected area	**Basal stem rot** (see page 19)
Roots in the garden	– split	**12**
	– forked	**8**
	– hollowed out	**13**
	– green-topped	**2**
	– small	**5**
	– covered with purple mould	**6**
	– covered with white mould	**7**
	– scurfy black patches	**4**
	– black patches, decay inside	**11**
	– tunnelled	**1** or **Wireworm** (see page 5)
	– eaten	**Cutworm** (see page 5) or **Millepede** (see page 5), or **Slugs & Snails** (see page 5)
Roots in store	– covered with purple mould	**6**
	– covered with white mould	**7**
	– sunken black areas	**3**
	– soft, evil-smelling	**Soft rot** (see page 26)

Carrots are not considered easy vegetables to grow successfully, and in some areas carrot fly attacks reach such proportions that it is common practice to delay seed sowing until mid July. Success depends mainly upon your soil type. If it is very heavy and sticky then healthy, large-sized roots are virtually an impossibility; so choose a short-rooted variety if you have had problems in the past. If you have failed with parsnips then choose a short-rooted canker resistant variety.

TUNNELLED ROOTS

¼ in. creamy maggots

1 CARROT FLY

Tell-tale signs are reddish leaves which wilt in sunny weather. At a later stage the leaves turn yellow. This pest is the major disorder of carrots and parsnips. Seedlings are killed; mature roots are riddled and rendered susceptible to disease. Attacks are worst in dry soils.

Treatment: None.

Prevention: Grow carrots well away from tall plants. Sow thinly and destroy all thinnings. Rake Bromophos into the soil before sowing or apply along the seed drill. If crops are not to be lifted until the autumn, water around plants with spray-strength Crop Saver in August.

Golden Rules for staying out of trouble

Leave at least one season between manuring and sowing. Prepare the seed bed thoroughly. Rake Bromophos into the soil before sowing. Sow seed very thinly. Store only undamaged roots.

2 GREEN TOP

The tops of carrot roots are sometimes found to be green when the crop is harvested. Unlike potatoes green carrots are not harmful but they are unsightly. Green top is caused by sunlight on the exposed crowns, and it is easily prevented by earthing-up to cover the tops of the roots during the growing season.

5 SMALL ROOTS

Aphids and virus will stunt grow and reduce yield, but even in t absence of pests and diseases ma gardeners produce disappointing small carrots. Poor soil conditions a usually to blame – you must dig t soil deeply and break up clays by ad ing well-rotted compost or manure least one or two seasons before gro ing carrots. If the plants are slo growing, spray with a foliar fe (Fillip) at regular intervals or wat with Bio Plant Food. Carrots respo well to this in-season feeding.

3 BLACK ROT

A storage disease of carrots, which renders the root useless. The large black lesions are easily seen in store, but there are no symptoms on the growing crop.

Treatment: Burn diseased roots immediately.

Prevention: Store roots properly. Do not use the land for carrots next year.

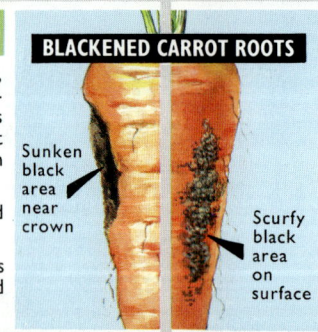

BLACKENED CARROT ROOTS

Sunken black area near crown

Scurfy black area on surface

4 CLAYBURN

This carrot disorder is neither common nor serious, and is always associated with pockets of clay in a loamy soil. Harmful salts contained in the clay cause the damage, but the culinary value is not affected.

Treatment: None.

Prevention: Dig out clay from soil used for growing carrots if you are growing for exhibition.

MOULDY ROOTS

MOULDY ROOTS

FORKED ROOTS

6 VIOLET ROOT ROT

An occasional disease of carrots and parsnips. The only above-ground symptom is a slight yellowing of the foliage, but harvested roots show a felt-like mass of purplish threads covering the lower parts.

Treatment: None. Destroy all diseased roots.

Prevention: Never store any roots which are affected by violet root rot or the whole crop will be lost. Do not grow root crops or asparagus on land affected in the previous season.

7 SCLEROTINIA ROT

The major disease of carrots in store. White woolly mould covers and soon destroys the roots. Occasionally this disease attacks the growing crop – lower leaf stalks and crown turn black.

Treatment: None. If attacks are seen in the garden, water healthy plants with Cheshunt Compound after lifting and destroying diseased plants. Remove rotten roots immediately.

Prevention: Dust seed with Benlate before sowing; store only sound roots in a dry, airy place. Do not grow carrots or celery on land affected by sclerotinia rot in the previous season.

8 FANGING

Fanging is usually caused by adding manure or compost to the soil shortly before seed sowing. Other causes are growing carrots in stony soil or in heavy ground which has not been properly dug.

Treatment: None.

Prevention: Use land which has been manured for a previous crop. Don't make the seed bed too firm.

DISCOLOURED LEAVES

DISTORTED LEAVES

BLACKENED PARSNIP ROOTS

black area at top of root

9 MOTLEY DWARF VIRUS

Central leaves show a distinct yellow mottling, outer leaves have a reddish tinge. This virus is spread by the carrot-willow aphid. The growth of infected plants is greatly reduced and the yield is small if the plants are attacked at the seedling stage.

Treatment: None.

Prevention: Keep young carrots free from aphids by spraying with Topgard Systemic Liquid, Crop Saver or Fenitrothion.

10 CARROT-WILLOW APHID

Greenfly attacks can be serious in a warm, dry summer. The leaves are distorted, discoloured and stunted. Plants are weakened, but even more serious is the transmission of motley dwarf virus by this pest.

Treatment: Spray at the first sign of attack with Topgard Systemic Liquid, Crop Saver or Fenitrothion.

Prevention: No practical method available.

11 PARSNIP CANKER

A serious disease of parsnips, which can be caused by several factors: soil acidity, presence of fresh organic matter in soil, root damage and irregular rainfall. The blackened areas on the roots crack and the parsnips rot.

Treatment: None.

Prevention: Lime soil. Don't sow too early. Do not grow a susceptible variety on the same site next year; choose a resistant variety such as Avonresister or White Gem.

12 SPLITTING

Much more serious than the fanging of carrots, because these roots will not store. The cause is heavy rain or copious watering after a prolonged dry spell.

Treatment: None. Use split roots immediately.

Prevention: Water thoroughly in times of drought. Apply a mulch of peat or compost around the crop in dry weather.

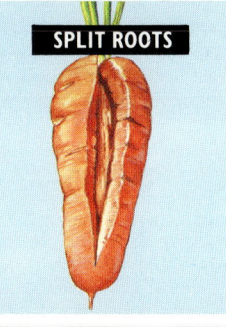

SPLIT ROOTS

HOLLOWED-OUT ROOTS

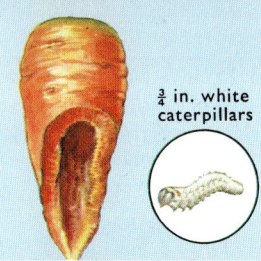

¾ in. white caterpillars

13 SWIFT MOTH

These soil-living caterpillars, which move backwards when disturbed, hollow out the roots of carrots and parsnips.

Treatment: None. Burn affected roots. Destroy caterpillars.

Prevention: Controlled by Bromophos used for carrot fly prevention.

Celery

WHAT'S WRONG...

	Symptom	Likely Causes
Seedlings	– toppled over	Damping off (see page 5) or Basal stem rot (see page 19)
Leaves	– tunnelled, blistered	1
	– covered with brown spots	2
	– yellow, withered	Boron deficiency (see page 4)
	– yellow, not withered	Cucumber mosaic virus (see page 17)
Stalks	– tough, bitter	1
	– pithy, not crisp	Dry soil
	– split vertically	5
	– split horizontally	Boron deficiency (see page 4)
	– mouldy at base	Sclerotinia rot (see page 15)
	– eaten above ground level	Slugs & Snails (see page 5)
	– eaten at ground level	Cutworm (see page 5)
Hearts	– missing, flower stalk only	3
	– rotten	4
Roots	– eaten	Carrot fly (see page 14)

Celery is not an easy crop to grow successfully, and its culture is made even more difficult by four serious problems which can plague this crop. Three of these problems are easily noticed whenever they occur – celery fly, celery leaf spot and slugs. The fourth problem is shortage of water, and here the effects are less obvious but no less devastating. Prolonged dryness at the roots will invariably lead to the production of plants with inedible hearts.

BLISTERED LEAVES

1 CELERY FLY (Leaf Miner)

White $\frac{1}{4}$ in. maggots tunnel within the leaves, causing blisters to develop. Attacks occur from May onwards, and the effects are most serious on young plants. Whole leaves may shrivel and die, and the stalks are stunted and bitter.

Treatment: Pinch out and destroy affected leaflets. Spray with Malathion or Crop Saver at the first sign of attack.

Prevention: Never plant seedlings with blistered leaves.

2 CELERY LEAF SPOT (Blight)

Brown spots appear first on the outer leaflets and then spread to all of the foliage. In a wet season the whole plant may be destroyed if the disease is not checked.

Treatment: Spray with Benlate at the first sign of disease. Repeat as necessary.

Prevention: Buy seed described as 'thiram treated' or 'hot water treated'. Never plant seedlings with spotted leaves.

SPOTTED LEAVES

3 BOLTING

Bolting is a serious problem, which unfortunately is common in dry seasons. At lifting time the heart is found to contain just one inedible flower stalk instead of the expected cluster of edible stalks. There are several possible causes, the most likely being dry soil conditions around the roots. Never let plants go thirsty during drought. Bolting can also be caused by planting out seedlings which have grown too large or have been damaged by frost.

ROTTEN HEARTS

4 CELERY HEART ROT

This disorder is noticed at lifting time. On cutting the plant open, the heart is found to be a slimy brown mass. The bacteria which cause the rot enter the stalks through wounds caused by slugs, frost or careless cultivation.

Treatment: None. Destroy diseased plants.

Prevention: Grow celery on well-drained land. Keep slugs under control and take care when earthing-up. Bacteria build up in the soil after an attack so do not grow celery on land affected in the previous season.

SPLIT STALKS

5 SPLITTING

Celery stalks are sometimes spoilt by vertical splitting. This disorder is usually caused by dry soil around the roots, but it can be due to an excess of nitrogen in the soil.

Treatment: None.

Prevention: Water thoroughly in dry weather, especially during the early stages of growth. Feed regularly with a liquid fertilizer, such as Bio Plant Food, which contains more potash than nitrogen.

Golden Rules for staying out of trouble

Grow in good soil. Sow treated seed. Scatter Slug Pellets around plants. Never let the soil dry out and earth up carefully. Feed with Bio Plant Food during summer months.

The Cucumber Family
CUCUMBER · MELON · COURGETTE · MARROW

Greenhouse cucumbers are a delicate crop, and a host of bacterial and fungal infections can attack them. Most of these troubles arise through incorrect soil preparation or careless management of the growing plants, so study page **16** of the Vegetable Plotter if you are a beginner. Outdoor cucumbers and marrows are much simpler to grow and are generally trouble-free, although slugs, grey mould, powdery mildew and cucumber mosaic virus can cause serious losses.

WHAT'S WRONG...

	Symptom	Likely Causes
Seedlings	— eaten	**10** or **Woodlice** (see page 5) or **Flea beetle** (see page 12)
	— toppled over	**Damping off** (see page 5)
Stems	— gnawed at base	**Millepede** (see page 5)
	— brown shrivelled patches	**18**
	— mouldy patches	**6** or **16**
	— soft brown rot at base	**19**
Leaves	— holed	**10** or **Woodlice** (see page 5)
	— wilted	**1** or **4** or **5** or **18** or **19** or **20**
	— yellow, moving up plant	**20**
	— mottled yellow and green	**1**
	— covered with silky webbing	**21**
	— covered with spots	**16** or **17**
	— covered with mould	**6** or **22**
	— infested with greenfly	**Aphid** (see page 5)
	— tiny moths, sticky surface	**Greenhouse whitefly** (see page 29)
	— papery patches	**23**
	— brown patches, yellow halo	**18**
Roots	— blackened, rotten	**5**
	— covered with galls	**4**
Fruit	— no flowers	**Lack of humidity**
	— no fruit	**2**
	— covered with mould	**6** or **12**
	— sunken spots	**7** or **9**
	— tip rotten, oozing gum	**13**
	— eaten	**10** or **11**
	— young fruits withered	**14**
	— bitter	**15**
	— poor yield	**3**

MOTTLED LEAVES

1 CUCUMBER MOSAIC VIRUS

Cucumber mosaic virus is a common and extremely serious disease. Marrows are even more susceptible than cucumbers. The leaves are mottled with yellow and dark green patches. The leaf surface becomes puckered and distorted. Plants are severely stunted and may collapse in a bad attack.

Treatment: None. Destroy all infected plant material; wash hands and tools thoroughly before touching other plants.

Prevention: This disease is spread by greenfly, so spray immediately with Crop Saver if these pests are seen.

2 NO FRUIT

A common complaint of marrows and courgettes is the failure of fruit to set. The usual cause is poor pollination and it is wise to give nature a helping hand. This calls for fertilizing 2 or 3 female flowers (tiny marrow behind petals) by dusting a male flower (thin stalk behind petals) into the mouth of each one. This job should be done in the morning, preferably on a dry day. Make sure the soil is kept moist.

3 POOR YIELD

Greenhouse cucumbers sometimes lose their vigour shortly after the first fruits have been picked. To keep the plants cropping it is necessary to follow a few simple rules. Remove the first fruits when they are small. Encourage root activity by adding a mulch around the stems. Feed every 2 weeks with Bio Plant Food. Cut the fruits when they have reached a reasonable size; if the fruits mature then flower production will cease.

4 EELWORM

Both indoor and outdoor crops may be attacked by root knot eelworm. Gall-like growths develop on the roots. Leaves are discoloured.

Treatment: None. Lift and destroy badly wilted plants.

Prevention: Do not grow cucumbers in infested soil for at least 6 years.

ROOT TROUBLES

5 ROOT ROT

Several fungal diseases can affect the root system; black root rot is the worst. The tap root becomes black and rotten causing the plant to wilt.

Treatment: None. Lift and destroy collapsed plants.

Prevention: Grow plants in sterilized soil. Avoid cold and overwet growing conditions.

Golden Rules for staying out of trouble

Use compost or sterilized soil under glass. Pay careful attention to ventilation and watering. Feed regularly once fruit appear. Mulch stem bases with moist peat. Remove diseased leaves; pick fruit regularly.

The Cucumber Family continued

FRUIT TROUBLES

6 GREY MOULD (Botrytis)

A grey furry mould appears on rotting fruit. Botrytis can cause serious losses outdoors in a wet season and under glass if the humidity is high. Stems are frequently infected, the point of entry being a damaged or dead area.

Treatment: Remove and burn infected fruit and leaves. Spray with Benlate at the first signs of disease.

Prevention: Avoid overwatering. Spray plants regularly with Benlate.

7 GUMMOSIS

A serious disease of greenhouse or frame cucumbers grown under wet and cool conditions. Infected fruits develop sunken spots through which oozes an amber-like gum. A dark mould develops on the surface of this gum.

Treatment: Destroy all diseased fruit. Raise the temperature and reduce the humidity. Spray the plants with Dithane.

Prevention: Keep the greenhouse or frame warm and ensure adequate ventilation.

8 CUCUMBER MOSAIC VIRUS

Misshapen small fruits bearing distinctive dark green warts. The surface is either white or yellow with patches or spots of green. The severity of the symptoms increases with the temperature of the greenhouse.

Treatment: None. Healthy plants should not be handled after infected fruit have been cut. However, there is no health risk if virus-affected cucumbers or marrows are eaten.

Prevention: See page 17.

9 ANTHRACNOSE

Pale green sunken spots and patches appear near the blossom end of the fruits. The affected areas turn pink as mould develops over the surface, and eventually they become black and powdery. As the disease spreads the affected fruits turn yellow and die.

Treatment: None. Destroy infected fruit and dust plants weekly with sulphur.

Prevention: Sterilize the soil before growing cucumbers next year. Make sure the greenhouse is adequately ventilated.

10 SLUGS & SNAILS

As marrows increase in size they become susceptible to attack by slugs and snails. The outer layers are scraped away and the soft flesh is then eaten.

Treatment: Scatter Slug Gard or Slug Pellets around the plants at the first signs of attack.

Prevention: Keep area free from rubbish. Place a slate or tile beneath each growing fruit.

11 MICE

Mice are attracted by the smell of ripening melons and they can cause serious losses of frame-grown crops.

Treatment: None.

Prevention: Keeping mice away from frame-grown melons is difficult if there is a nest nearby. The pests should be destroyed before the fruit mature by using Racumin. In a greenhouse the fruit can be kept away from mice by supporting them in string bags.

12 SCLEROTINIA ROT

Dark rotten areas on greenhouse cucumbers develop white cottony mould. In this mould large black cyst-like bodies are formed.

Treatment: None. Pick and destroy infected fruit immediately.

Prevention: Avoid splashing the fruit during watering and prevent them from coming into contact with soil.

13 BLACK ROT

Rotting of the ends of cucumbers, and the oozing of gum in the shrivelled diseased area, indicates attack by the fungus which causes stem rot (see page 19).

Treatment: None. Pick and destroy infected fruit immediately. Stop spraying temporarily.

Prevention: Sterilize the soil before growing cucumbers next year.

14 WITHERING OF YOUNG FRUITS

Cucumbers and marrows stop growing when they are only a few inches long and withering spreads back from the tip. Unfortunately there are many possible causes, such as draughts, heavy pruning and the use of fresh farmyard manure. The most likely reason is faulty root action due to poor drainage, overwatering or poor soil preparation. The secret is to maintain steady growth by careful watering. If withering of young fruits does take place, remove the damaged fruits and spray with Fillip. For the next week withhold water and ventilate the greenhouse. Syringe as usual.

15 BITTERNESS

If the fruit is normal in appearance then one of the growing conditions is at fault. A sudden drop in temperature or soil moisture and a sudden increase in sunshine or pruning are all common causes. The second type of bitterness is associated with misshapen club-like fruits grown under glass. Here the cause is pollination; remember that male flowers must be removed. This tedious job can be avoided by growing an all-female variety such as Femdam. Bitter cucumbers are generally unusable, but you can try the old practice of cutting the fruit a couple of inches from the blossom end and rubbing the cut surfaces together.

SPOTS ON LEAVES

SHRIVELLED BLOTCHES

16 ANTHRACNOSE (Leaf Spot)

Small pale spots rapidly enlarge and turn brown. Each spot has a yellowish margin. In a bad attack the spots fuse and the leaf withers. Large areas of pink mould develop on the stems and leaf stalks. These areas eventually turn black.

Treatment: Remove and burn spotted leaves. Dust weekly with sulphur. Lift and burn badly diseased plants.

Prevention: Sterilize the soil before growing cucumbers next year. Make sure the greenhouse is adequately ventilated.

17 BLOTCH

Less common than Anthracnose, and the spots are usually smaller and paler. Leaves decay rapidly in a severe attack. Unlike Anthracnose, the disease does not affect the stems, and pink mould does not develop.

Treatment: Remove and burn spotted leaves. Dust weekly with sulphur. Lift and burn badly diseased plants.

Prevention: Sterilize the soil before growing cucumbers next year. Alternatively grow 'Butcher's Disease Resisting'.

18 STEM ROT

Leaf blotches have a distinct yellow halo. Affected area turns brown and shrivelled. Stems are attacked (gummy stem blight) and may be killed. Fruit are also affected (black rot, see page 18).

Treatment: Remove and burn all diseased plant material. Stop damping down temporarily.

Prevention: Sterilize the soil before growing cucumbers next year.

COLLAPSED STEMS

YELLOWING LEAVES

SILKY WEBBING

19 BASAL STEM ROT

This bacterial disease has several common names, including soft rot, cucumber foot rot and canker. The brown slimy rot attacks the base of the stems of greenhouse crops. The leaves wilt and the plant may collapse.

Treatment: If the plant is not too badly damaged dust sulphur over the brown area and then apply a moist peat mulch around the stem to cover the diseased zone. If attack is severe then lift and destroy affected plants.

Prevention: Avoid overwatering and keep water away from the base of the stem.

20 VERTICILLIUM WILT

Lower leaves turn yellow and the discoloration moves upwards. Finally all leaves become dry and wilted. Tell-tale signs are brown streaks inside the stem tissue (see page 29). Most susceptible are young plants growing in cold, wet conditions.

Treatment: Keep air moist and warm. Shade the greenhouse and do not overwater.

Prevention: Sterilize soil before growing cucumbers or tomatoes.

21 RED SPIDER MITE

Fine silky webbing occurs over the leaves and stems. The foliage appears speckled and bleached, and the tiny mites can be found on the underside. Growth is retarded and the shoots are thin and weak. The mites are greenish in summer, bright red in winter.

Treatment: Spray with Derris or Malathion at the first signs of attack.

Prevention: Maintain a damp atmosphere in the greenhouse.

22 POWDERY MILDEW

Leaves and stems are covered with white powdery patches. This disease occurs outdoors in a warm dry summer and under glass. It is encouraged by dry soil combined with a moist atmosphere.

Treatment: Spray with Benlate at the first signs of disease.

Prevention: Keep the soil moist at all times and ventilate adequately under glass.

WHITE POWDERY MOULD

PAPERY PATCHES

23 SUN SCALD

Pale brown, papery patches sometimes occur on the margins of the leaves and shrinkage of the dry areas occurs. Exposure to bright sunlight is the cause.

Treatment: None.

Prevention: Paint glass with Coolglass. Damp down adequately, but not at midday as water droplets can act as magnifying lenses.

Lettuce

WHAT'S WRONG...

Outdoor lettuce is an easy crop to grow, but it is not easy to grow well. You must guard against soil pests, slugs and birds; in cool, damp weather the twin major diseases (powdery mildew and grey mould) can be destructive. Above all you must try to prevent any check to growth. Crops grown under glass are vulnerable to an even wider range of plant troubles, but few of them are serious in a well-grown crop.

	Symptom	Likely Causes
Seedlings	— poor or slow germination	**Seeds kept too warm**
	— eaten	**5** or **10** or **Birds** (see page 8) or **Mice** (see page 6) or **Millepede** or **Leatherjacket** (see page 5)
	— toppled over	**9** or **Damping off** (see page 5)
	— severed	**10**
Leaves	— holed	**4** or **5**
	— mouldy or powdery patches	**1** or **9**
	— brown-edged	**1** or **9** or **13**
	— large yellowish patches	**1**
	— brown spots	**4**
	— mottled	**12**
	— infested with greenfly	**8**
Plants	— run to seed	**2**
	— no hearts	**3**
	— wilted	**6** or **7** or **9** or **10** or **11**
	— rotten at base	**9**
	— base covered with grey mould	**9**
	— base covered with fluffy white mould	**Sclerotinia rot** (see page 15)
Roots	— eaten	**10** or **Leatherjacket** or **Millepede** (see page 5)
	— tunnelled	**11**
	— infested with greenfly	**6**
	— covered with white patches	**6**
	— covered with gall-like swellings	**7**

WHITE MOULDY LEAVES

1 DOWNY MILDEW

Large yellowish patches appear between the veins of older leaves. Whitish mouldy areas develop on the underside. Later, diseased patches turn brown and die. This serious disease is worst in cool, wet conditions.

Treatment: Remove affected leaves as soon as they are seen and spray the plants with Dithane.

Prevention: Practise crop rotation. Avoid overcrowding. Under glass make sure that the plants are adequately ventilated and not overwatered.

2 BOLTING

Lettuces produce thick flowering stems if left in the soil after hearts have formed. Sometimes the plants run to seed before they are ready for harvesting; this condition is known as 'bolting'. The cause is a check to growth at some stage of the plant's life. Careless or delayed transplanting is perhaps the commonest cause, but both over-crowding and dryness at the root are frequently responsible. Lift and place on the compost heap; cover with soil so as not to attract aphids.

3 NO HEARTS

A wide variety of factors can prevent lettuces from forming hearts. The most likely reason is shortage of organic matter; you must enrich the land with compost or manure if you want to be sure of a well-hearted crop. Other possible causes are growing the plants in a shady site, aphid attack, overcrowding and drought.

4 RING SPOT (Rust)

Not common, but occasionally damaging to winter varieties. Small brown spots appear on the outer leaves, giving a rusty appearance. The centres of the spots may fall out. Rusty streaks appear on the leaf midribs.

Treatment: Destroy badly infected plants. Spray remaining plants with Hexyl Plus.

Prevention: Practise crop rotation outdoors. Ensure good ventilation under glass.

HOLED LEAVES

5 SLUGS & SNAILS

Both slugs and snails are a menace to lettuces at all stages of growth. Seedlings are particularly susceptible and may be killed. Leaves and stems are severely attacked in wet weather. The pests are generally not seen during the day, so look for tell-tale slime trails.

Treatment: Scatter Slug Gard or Slug Pellets around the plants at the first signs of attack.

Prevention: Keep surrounding area free from rubbish.

Golden Rules for staying out of trouble
Enrich soil with compost or manure. Rake in Bromophos. Choose a suitable variety for time of sowing. Sow seed where the plants are to grow and mature. Feed and water regularly.

DAMAGED ROOTS

6 ROOT APHID

Greyish-coloured 'greenfly' attack the roots, which become covered with white powdery patches. Growth is stunted, and the leaves may turn yellow and wilt. Attacks are worst in late summer.

Treatment: Pull up and destroy decaying plants. Water around remaining plants with spray-strength Malathion.

Prevention: Keep plants watered in dry weather. Grow resistant varieties (Salad Bowl, Avoncrisp) if attacks reoccur.

DAMAGED ROOTS

7 EELWORM

Root knot eelworm occasionally attacks lettuce plants, causing stunted growth and pale-coloured leaves. Plants wilt and die if the attack is severe. Look for the tell-tale sign on lifted plants – gall-like swellings on the roots.

Treatment: None. Dig up and destroy infested plants.

Prevention: Do not grow lettuce in affected soil for at least 6 years.

GREENFLY ON LEAVES

8 APHID

Greenflies can be serious pests in two ways. They spread mosaic, a virus disease, and they also cover the plants in sticky honeydew which can make them unusable. Attacks are worst in a dry spring, when leaves may be badly puckered and distorted.

Treatment: Spray at the first signs of attack. Use Topgard Systemic Liquid, or Sprayday if plants are ready for cutting.

Prevention: None.

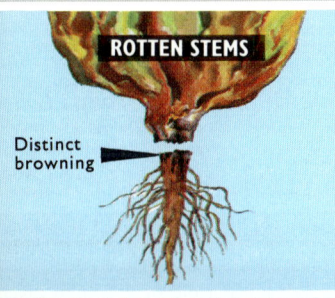

ROTTEN STEMS

Distinct browning

9 GREY MOULD (Botrytis)

Plants are infected through dead or damaged areas, and the fungus produces a reddish-brown rot when it reaches the stem. Plants wilt and may break off at soil level. Infected tissue produces abundant masses of grey mould. This serious disease is encouraged by low temperature and high humidity.

Treatment: Destroy diseased plants immediately. Spray thoroughly with Benlate.

Prevention: Handle seedlings carefully. Plant so that the leaf bases are not buried.

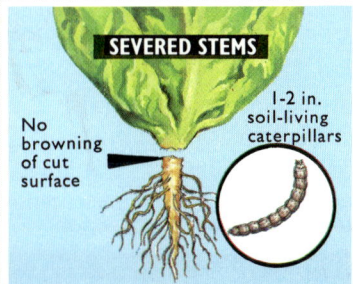

SEVERED STEMS

No browning of cut surface

1-2 in. soil-living caterpillars

10 CUTWORM

These large caterpillars are a major threat to young lettuces. The plants are attacked at night and stems may be severed at ground level. With older plants the roots are gnawed which causes the lettuces to wilt. June and July are the danger months.

Treatment: Hoe the soil around the plants. Destroy caterpillars which are brought to the surface.

Prevention: Rake Bromophos into the soil before planting.

TUNNELLED ROOTS

¼ in. maggots

11 LETTUCE ROOT MAGGOT

An occasional pest of lettuce grown under glass. They tunnel into the roots and eat the central tissue. Above ground the leaves wilt and growth is stunted. The main host of this pest is the chrysanthemum, where it is known as chrysanthemum stool miner.

Treatment: Dig out and destroy affected plants. Water around remaining plants with spray-strength Hexyl Plus.

Prevention: Do not plant lettuce on land used for chrysanthemums in the previous season.

12 MOSAIC

Yellow or pale-green mottling appears on the leaves and growth is stunted. The veins appear almost transparent. This virus disease is spread by aphids.

Treatment: None. Lift and burn infected plants.

Prevention: Spray young plants with Topgard Systemic Liquid or Crop Saver.

MOTTLED LEAVES

BROWN-EDGED LEAVES

13 TIPBURN

Tipburn (or 'greasiness') is a common cause of the scorching of leaf edges which sometimes occurs. No mould is present, and it is usually due to sudden water loss by the leaves. This can happen in a warm spell in early spring or at the start of a summer heat wave.

Treatment: None.

Prevention: None.

Onions & Leeks

WHAT'S WRONG...

	Symptom	Likely Causes
Seedlings	– eaten	**Cutworm** (see page 5)
	– killed	**4**
	– toppled over	**Damping off** (see page 5)
Sets	– lifted out of ground	**Frost** or **Birds**
	– two or more plants produced	**6**
Leaves	– tunnelled	**12**
	– eaten above ground level	**Cabbage moth** (see page 11)
	– eaten at ground level	**Cutworm** (see page 5) or **Wireworm** (see page 5)
	– diseased	**7** or **8** or **11**
	– yellow, drooping	**1** or **13** or **14**
	– green, drooping	**2**
	– white tipped	**10**
Plants	– swollen, distorted	**4**
	– abnormally thick-necked	**9**
	– run to seed	**3**
Bulbs in the garden	– tunnelled, maggots present	**1**
	– split at base	**5**
	– mouldy at base	**13**
	– soft, not evil-smelling	**4** or **11** or **13**
	– soft, evil-smelling inside	**14**
Bulbs in store	– soft, mouldy near neck	**15**
	– soft, evil-smelling	**Soft rot** (see page 26)

Although many plant disorders can attack onions, only four are likely to seriously trouble the gardener. They are onion fly, stem and bulb eelworm, neck rot and white rot. Plants grown from seed are more susceptible to onion fly, so raise onions from sets if you have been disappointed in the past. Leeks are much less prone to attack than onions.

TUNNELLED BULBS

¼ in. white maggots

1 ONION FLY

Tell-tale signs are yellow drooping leaves. Worst attacks occur in dry soil in midsummer. The maggots burrow into the bases of the bulbs; young plants are frequently killed, older ones fail to develop properly.

Treatment: Lift and burn badly affected plants.

Prevention: Destroy all thinnings and damaged leaves; firm the soil around the plants. Apply Calomel Dust (4 oz per 15 yards of row) around seedlings when they are 1 inch high. Alternatively rake Bromophos into the soil shortly before sowing or planting.

2 DROOPING LEAVES

Leaves sometimes droop even though neither pest nor disease is present. If foliage is darker green than normal, then the usual cause is either too much fresh manure before planting or too much nitrogen in the soil. Watering with a potash-rich fertilizer, such as Bio Tomato Food, will help.

3 BOLTING

Onions occasionally bolt (premature production of flower heads). When this happens cut off the flower stalks and lift bulbs in the usual way. Use as soon as possible as they will not store satisfactorily. Common causes are early sowing or planting in a cold spring and planting in loose soil.

TWISTED, SWOLLEN LEAVES

4 STEM & BULB EELWORM

Swollen, distorted foliage indicates attack by this microscopic soil-living pest. Young plants are killed; older ones produce soft bulbs which cannot be stored.

Treatment: Lift and burn infected plants.

Prevention: Do not grow onions, peas, beans or strawberries on land affected by stem and bulb eelworm for several years.

BULBS SPLIT AT BASE

5 SADDLEBACK

Harvested onions are found to be split at the base. This disorder affects crops grown from sets, and it is always associated with heavy rain or watering after a prolonged period of drought.

Treatment: None. Use affected onions as soon as possible after harvesting as they will not keep in store.

Prevention: Never keep the plants short of water during dry spells in summer.

Golden Rules for staying out of trouble

Buy good quality seed or sets. Choose a fresh site each year. Don't leave thinnings and damaged foliage in the garden. Store only hard bulbs and keep them cool.

6 SET DIVISION

Onions grown from sets may produce twin bulbs. The cause of this splitting is usually planting at the wrong time or growing the plants in poor soil. Prolonged dry weather can also induce set division.

7 SMUT

Black spots and blotches appear on leaves and bulbs. Only young plants are infected; the infected leaves become thickened and twisted. Leeks are more susceptible than onions.

Treatment: None. Lift and burn diseased plants.

Prevention: Do not grow onions or leeks on infected land for at least 8 years.

SPOTS ON LEAVES

8 RUST

Orange spots and blotches appear on the surface of leaves. Uncommon, but effect can be fatal in a severe attack in summer. Leeks are more susceptible than onions.

Treatment: Remove and burn diseased leaves.

Prevention: Do not grow leeks or onions on land affected by rust in the previous season.

WHITENED LEAVES

9 BULL NECK (Thick Neck)

The production of abnormally thick necks is a serious complaint as the bulbs will not store properly. Bull-necked onions are often associated with the overmanuring of land and the use of too much nitrogen. Use a liquid feed during the growing season; choose one such as Bio Plant Food which contains more potash than nitrogen. Another possible cause is sowing seed too deeply.

MOULD ON LEAVES

10 WHITE TIP

Leaf tips of leeks turn white and papery in late summer or autumn. Disease spreads downwards and growth is stunted.

Treatment: Spray with Dithane as soon as the first signs are noticed. Lift and burn badly diseased plants.

Prevention: Do not grow leeks on land affected in the previous season.

11 DOWNY MILDEW

Downy grey mould covers leaves, which slowly die back and shrivel from the tips. Bulbs are usually soft and not suitable for storage. A serious disease in cool, damp seasons.

Treatment: Spray with Dithane at the first signs of disease. Repeat every fortnight.

Prevention: Grow onions on a different site each year. Avoid badly-drained areas.

TUNNELLED LEAVES

12 LEEK MOTH

Pale green $\frac{1}{2}$ in. caterpillars feed inside young onion leaves so that only outer skin remains. The foliage of leeks may also be attacked.

Treatment: Spray with Crop Saver or Fenitrothion at the first signs of attack. Destroy badly affected leaves.

Prevention: No practical method available.

ROTTING BULBS IN STORE

ROTTING BULBS IN THE GARDEN

13 WHITE ROT (Mouldy Nose)

Foliage turns yellow and wilts. Fluffy white mould appears on the base of the bulbs, and round black bodies appear in this fungus. White rot is a serious disease, and is always worst in hot, dry summers.

Treatment: None. Lift and burn diseased plants.

Prevention: Dust the seed drills with Calomel Dust. Do not grow onions on infected land for at least 8 years.

14 SHANKING

Centre leaves turn yellow and collapse; outside leaves soon follow. Cut open a bulb; the tell-tale sign is evil-smelling slime within the scales. This disease is much less common than white rot.

Treatment: None. Lift and burn diseased plants.

Prevention: No practical method available. Do not grow onions on infected land for several years.

15 NECK ROT

Grey mould appears near neck in store; bulbs turn soft and rotten.

Treatment: None. Examine stored bulbs frequently and remove rotten bulbs immediately.

Prevention: Dust seeds and sets with Benlate before planting. Follow all the rules for correct storage—dry bulbs thoroughly; store only hard, un-damaged ones in a cool well-ventilated place. Don't store onions with fleshy, green necks.

Potatoes

WHAT'S WRONG...

	Symptom	Likely causes
Seed potatoes	– thin, long sprouts	2
	– no sprouts	11
Leaves	– pale green or yellow	6 or 12 or 14 or **Drought**
	– pale green or yellow mottling	4 or 9
	– yellow or brown between veins	10
	– rolled, brittle	3
	– torn	9
	– brown patches at margins	1
	– many small holes	7
	– tiny brown spots	7
	– greenfly clusters	6
Stems	– tunnelled	13
	– blackened at base	12
Roots	– covered with tiny cysts	14
Tubers	– abnormally small	3 or 4 or 12 or 14 or **Drought**
	– contain narrow tunnels	21
	– contain wide tunnels	15 or **Millepede** (see page 5) or **Cutworm** (see page 5)
	– hollowed out	23
	– split	22
	– scabby spots on surface	16 or 18
	– wrinkled area, small woolly growths	20
	– brown lines in flesh	17
	– brown areas under skin	19
	– flesh slimy, evil-smelling	25
	– sunken brown area on surface	26
	– cauliflower-like warty outgrowths	24
	– black at centre	8
	– poor flavour or texture	8
	– soft at lifting time	5

Many diseases, pests and disorders can attack potatoes and reduce yields, but only four are likely to be a serious threat. Three of these are pests – potato cyst eelworm, slugs and wireworm. The other one is a disease – potato blight. The virus diseases can be a menace and you should therefore buy potatoes which are Certified.

BROWN BLOTCHED LEAVES

1 POTATO BLIGHT

Blight is the most serious potato disease, capable of destroying all the foliage during August in a wet season. The first signs are brown patches on the leaves. Look on the underside of the leaflets – each blight spot has a white mould fringe in damp weather.

Treatment: None, once the disease has firmly taken hold.

Prevention: Plant healthy seed tubers. Spray with Dithane in July and repeat at fortnightly intervals if the weather is damp. If blight spots are already present, spraying will slow down the spread of the disease to other plants.

2 SPINDLY SPROUTS

By far the most common cause of spindly sprouts is keeping the tubers too dark or too warm prior to planting. If threadlike shoots form despite standing the tubers in a light, cool place then virus infection is a possible cause. Alternatively the tubers may have been slightly frosted. Always buy good quality seed potatoes and sprout them in a light, frost-free location.

5 SOFT TUBERS

Some of the tubers lifted after an extremely dry summer may appear to be perfectly sound on the outside but are soft and rubbery to the touch. This is not a disease; it is a disorder caused by the plant withdrawing water from the developing tubers. It can be prevented by watering thoroughly during drought.

ROLLED LEAFLETS

3 LEAF ROLL VIRUS

Leaf roll is one of the most serious virus diseases which attack potatoes. Leaflets roll upward and become hard and brittle. Affected plants are stunted and the yields are poor.

Treatment: None.

Prevention: Use certified seed. Spray with Crop Saver to control the virus-carrying aphids.

MOTTLED LEAVES

4 MOSAIC VIRUS

There are several mosaic diseases, and the symptoms vary with the potato variety grown. The usual tell-tale sign is yellow or pale green mottling over the whole leaf surface. Brown streaks may also appear.

Treatment: None.

Prevention: Use certified seed. Spray with Crop Saver to control the virus-carrying aphids.

6 APHID (Greenfly)

In a dry warm season the foliage may be heavily infested with greenfly. Plants are weakened and leaflets turn brown and may die. Most serious effect, however, is the spread of virus diseases by these sap-sucking pests.

Treatment: Spray with Crop Saver or Topgard Systemic Liquid.

Prevention: No practical method available.

INSECTS ON LEAVES

¼ in. greenish insects

7 CAPSID BUG

Small brown spots which later turn into holes appear on the foliage. Young shoots may be distorted and the crinkling of small leaflets may be severe.

Treatment: Damage is usually too slight to affect yield. Spray with Crop Saver or Fenitrothion if attack is severe.

Prevention: No practical method available.

Golden Rules for staying out of trouble

Buy certified seed. Do not lime but use a fertilizer. Pre-sprout tubers and plant only healthy ones. Earth up. Spray against blight in damp weather. Do not store damaged tubers. Choose a fresh site next year.

YELLOW-BLOTCHED LEAVES

9 FROST

Late spring frosts can severely damage the young shoots, delaying the development of early varieties. A severe frost can turn the stems black; less severe frosts cause yellow patches and torn leaflets on the older foliage.

Treatment: None.

Prevention: Cover young shoots of earlies with newspaper if frosts are expected.

8 POOR QUALITY

The cooking and eating qualities of tubers can sometimes be disappointing. Obviously all the tuber troubles described on the next page make preparation difficult or impossible, but several disorders do not show up until the potatoes are being cooked or eaten. A **soapy** or **waxy texture** is usually due to lifting the tubers before they are mature. It can also be caused by growing potatoes on chalky soil. A **sweet taste** is usually due to keeping the tubers too cold during storage. An **earthy taste** is caused by the presence of powdery scab or by growing the plants in BHC-treated soil. Potatoes sometimes have a **black heart** or turn **black when cooked.** The major causes are storage at over 100°F and potash deficiency.

11 GAPPING

The failure of seed potatoes to develop sprouts is usually due to the presence of disease in the tuber. Another possible reason for this failure is the frosting of the seed tubers in transit or during storage. If such faulty seed is planted then a gap will occur where a plant should be.

BROWNING BETWEEN VEINS

10 MAGNESIUM DEFICIENCY

The first symptom is a yellowing of the tissue between the veins of the leaflets. These yellow areas then turn brown and brittle. Growth is stunted.

Treatment: Apply a sequestered compound. Repeated spraying with Fillip may help.

Prevention: Incorporate a fertilizer, such as Crop Booster, which contains magnesium when preparing the soil for planting.

BLACKENED STEMS

12 BLACKLEG

Tell-tale sign is the blackening of the stems at and below ground level. The leaves turn yellow and wilt; eventually the haulm withers. This disease attacks early in the season and is worst in heavy soils and rainy weather

Treatment: None. Lift and burn affected plants.

Prevention: Never plant seed tubers which are soft and rotten. The practice of making seed potatoes go further by cutting them increases the risk of attack.

TUNNELLED STEMS

1 in. bristly caterpillars

13 ROSY RUSTIC MOTH

Potatoes grown in new gardens may have their stems hollowed out by these caterpillars. Affected plants die down earlier than normal.

Treatment: None. Dig out and destroy infested plants.

Prevention: No practical method available.

WITHERED LOWER LEAVES

Pinhead-sized cysts on roots

14 POTATO CYST EELWORM

Plants appear weak and stunted. Lower leaves wither away; upper leaves are pale green and wilt during the day. Haulm dies down prematurely. Marble-sized tubers are produced.

Treatment: None. Destroy infected plants and tubers.

Prevention: Practise crop rotation, especially in light soils. Do not grow potatoes or tomatoes on infected land for at least 6 years.

Potatoes continued

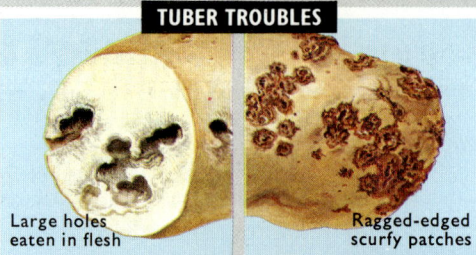

TUBER TROUBLES

15 SLUGS

Attacks begin in August; maincrop potatoes grown in heavy soil can be ruined.

Treatment: None.

Prevention: Avoid overmanuring. Apply Slug Gard in July. Lift the crop as soon as the tubers are mature.

Large holes eaten in flesh

Ragged-edged scurfy patches

16 COMMON SCAB

Disease is only skin deep; eating qualities unaffected. Most severe on light soils under dry conditions.

Treatment: None.

Prevention: Dig in compost but do not lime before planting. Use healthy seed.

17 SPRAING

Tubers are normal on the surface; discoloured inside. There are several causes – viruses, trace element deficiency and water shortage.

Treatment: None.

Prevention: Practise crop rotation. Do not grow Pentland Dell, which is very susceptible.

Curved lines in flesh

Round patches with raised margins

18 POWDERY SCAB

Much less frequent than common scab. Most severe on heavy soils under wet conditions. Scabs are powdery on the surface.

Treatment: None.

Prevention: Practise crop rotation. Do not grow Pentland Crown, which is very susceptible.

19 POTATO BLIGHT

Caused by blight spores from the leaves reaching the tubers. Affected potatoes rot in store.

Treatment: None. Do not store.

Prevention: Keep haulm well earthed up. If there is blight on the leaves then cut off and destroy stems 10 days before lifting.

Grey patches, reddish brown below skin

Shrunken area, whitish pustules

20 DRY ROT

Dry rot occurs in store. Careless handling when lifting makes tubers more susceptible.

Treatment: None. Destroy tubers.

Prevention: Practise crop rotation. Store only sound, healthy tubers and keep them in a cool but not over-dry atmosphere.

21 WIREWORM

Wireworm is a serious pest in new gardens, especially in wet summers. Tubers are riddled with narrow tunnels.

Treatment: None.

Prevention: Rake Bromophos into the soil before planting. Lift tubers as soon as they are mature.

1 in. orange shiny larvae

Deep cracks on surface

22 SPLITTING

Deep cracks make the tubers unattractive and difficult to peel. They are also very susceptible to rotting in store.

Treatment: None.

Prevention: Keep plants well watered during dry spells.

23 HOLLOW HEART

Affects large tubers. Caused by a prolonged wet spell after dry weather. Hollow-hearted potatoes may rot in store.

Treatment: None.

Prevention: Keep plants well watered during dry spells.

Hollow centre

Black warty outgrowths

24 WART DISEASE

Once very serious, now uncommon as all modern varieties are immune.

Treatment: None. Destroy affected tubers. Ministry of Agriculture must be informed of the outbreak.

Prevention: Plant only immune varieties on land known to be infected.

25 SOFT ROT

Soft rot is an infection which attacks damaged tubers. Affected tubers soon become slimy and putrid and are quite unusable.

Treatment: None. Destroy tubers.

Prevention: Store only sound, healthy tubers and make sure they do not become damp.

Soft, evil-smelling flesh

Dark brown depression on surface

26 GANGRENE

Gangrene occurs in store. Inside of tuber becomes decayed and hollow.

Treatment: None. Remove from store.

Prevention: Store only sound, healthy tubers and make sure the store is airy and frost-free.

Spinach

WHAT'S WRONG...

	Symptom	Likely Causes
Seedlings	— eaten	**Birds** or **Millepede** or **Slugs & Snails** (see page 5)
	— toppled over	**Damping off** (see page 5)
Leaves	— yellow between veins; acid soil	**Magnesium deficiency** (see page 13)
	— yellow between veins; chalky soil	**2**
	— holed	**Slugs & Snails** (see page 5)
	— spotted	**4**
	— infested with blackfly	**Black bean aphid** (see page 6)
	— infested with greenfly	**Aphid** (see page 5)
	— rolled	**5**
	— blistered	**Mangold fly** (see page 9)
	— yellow patches above	**1**
	— greyish-purple mould below	**1**
	— inner leaves narrow, yellow	**5**
Plants	— run to seed	**3**
	— early death, leaves deformed	**5**
	— early death, leaves not deformed	**Too hot and dry** or **Overcropping**

There are only three troubles which are likely to affect spinach, but they can make this a difficult crop to grow. Downy mildew, bolting and spinach blight are the major troubles, and if you have had problems with annual spinach in the past then try the much easier types — New Zealand spinach and spinach beet.

YELLOW PATCHES

1 DOWNY MILDEW

Watch for downy mildew if the weather is wet and cold. It begins on the outer leaves; yellow patches above and greyish-purple mould below. As the disease progresses affected patches turn brown.

Treatment: Pick off diseased leaves. Spray with Dithane at the first sign of attack.

Prevention: Practise crop rotation. Make sure the soil is well drained and avoid overcrowding by thinning the crop promptly.

2 MANGANESE DEFICIENCY

Yellow blotches appear between the veins, and the margins tend to curl up slightly. The symptoms are most pronounced in midsummer. Manganese deficiency is associated with chalky and sandy soils, which can make successful spinach growing difficult in such areas.

Treatment: Apply a sequestered compound. Repeated spraying with Fillip may help.

Prevention: Do not overlime the soil.

YELLOWED LEAVES

3 BOLTING

The commonest spinach trouble in home gardens is bolting, which results in the premature flowering of the plants. The danger is greatest in hot, settled weather and it will occur if the plants have been kept short of either water or nutrients. Avoid trouble by preparing the soil properly by digging in compost and raking in a vegetable fertilizer. Thin the seedlings early, and water in dry weather. In some soils bolting occurs year after year, and the best plan here is to grow New Zealand spinach.

SPOTTED LEAVES

4 LEAF SPOT

Numerous $\frac{1}{4}$ in. spots appear in the foliage; in a bad attack the spots join up and the leaf is destroyed. Central area of each spot is pale brown and may drop out; outer ring is dark brown or purple.

Treatment: Pick off and burn diseased leaves. Spray with Dithane.

Prevention: Practise crop rotation. Apply a balanced fertilizer, such as Crop Booster, before sowing seed.

ROLLED LEAVES

5 SPINACH BLIGHT

Young leaves are affected first. The telltale signs are narrow and small leaf blades, inrolled margins and a puckered, yellow surface. The cause of this serious disease is the cucumber mosaic virus.

Treatment: Destroy infected plants.

Prevention: Keep down weeds. Spray with Topgard Systemic Liquid or Crop Saver to control greenfly, which carry the virus.

Golden Rules for staying out of trouble

Dig in plenty of compost. Rake in fertilizer. Thin the seedlings as soon as possible. Water thoroughly in dry weather. Always cut (don't pull!) outer leaves. and harvest at frequent intervals.

Tomatoes

WHAT'S WRONG...

	Symptom	Likely causes
Seedlings	— eaten or severed	**Woodlice** or **Slugs** or **Cutworm** (see page 5)
	— toppled over	**Damping off** (see page 5)
	— gnawed roots	**Millepede** (see page 5)
Stems	— tunnelled	22 or **Wireworm** (see page 5)
	— grey mouldy patches	4
	— brown zone near soil level	6 or 7
Leaves	— blue tinged	**Too cold** or **too dry**
	— yellow between veins	12
	— grey mould	4
	— papery patches	17
	— brown patches	13
	— yellow patches on upper surface	3
	— mottled	1 or **Red spider mite** (see page 19)
	— curled	1 or 2 or 8 or 9
	— wilted	5 or 6 or 7 or 10 or 11
	— fern-like	1 or 8
	— infested with greenfly	**Aphid** (see page 5)
	— tiny moths, sticky surface	9
	— holed, caterpillars present	22
Roots	— brown, corky	5
	— covered with cysts or galls	10
Fruits	— drop before forming	16
	— form, but drop before maturity	4
	— form, but remain tiny	19
	— sticky, covered with black mould	9
	— soft rot	23
	— discoloured spots or patches	14 or 15 or 17 or 18 or 20 or 25
	— hollow	21
	— split or tunnelled	22 or 24

Diseases and disorders are much more important than insect pests – outdoor tomatoes are much less susceptible than crops grown under glass. Keep a careful watch and treat plants immediately symptoms appear. Tomatoes require regular feeding with a specific fertilizer, such as Bio Tomato Food, in order to prevent undersized fruit on the upper trusses. Don't overfeed – little and often is the secret.

DISTORTED OR DISCOLOURED LEAVES

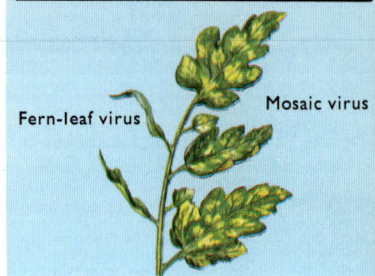

Fern-leaf virus   Mosaic virus

1   VIRUS

There are several important virus diseases which affect tomatoes. Leaves may be mottled and curled, stems may bear dark vertical streaks, foliage may be thin and distorted and growth may be stunted. Affected fruit is often mottled and bronzed.

Treatment: None. Destroy affected specimens. Feed remaining plants.

Prevention: Try to buy virus-free plants. Spray to control greenfly. Do not handle plants immediately after smoking.

2   LEAF ROLL

Unlike potatoes, rolled tomato leaves do not indicate disease. The inward curling of young leaves is usually taken as a good sign if they are dark green. The rolling of older leaves is usually due to excess deleafing or a wide variation between day and night temperatures. Provided that pests and disease are absent, there is no need to take remedial action.

5   ROOT ROT

Poor drainage can lead to root disease. Below ground the roots become brown and corky, above ground the plants tend to wilt in hot weather. Rots cannot be cured once they have taken hold; mulch around the stems with moist peat to promote the formation of new roots. Next year grow plants in bags, fresh compost or sterilized soil.

BROWN MOULD PATCHES

3   TOMATO LEAF MOULD

Purplish-brown mould patches appear on the underside of the foliage; the upper surface bears yellowish patches. Lower leaves are attacked first.

Treatment: Remove some of the lower leaves. Spray with Benlate at the first signs of attack.

Prevention: Ventilate the greenhouse, especially at night.

GREY FURRY PATCHES

4   GREY MOULD (Botrytis)

Grey mould usually starts on a damaged area of the stem. Other parts of the plant may then be infected – diseased flower stalks cause fruit drop.

Treatment: Cut out diseased areas and dust wound lightly with Benlate.

Prevention: Reduce humidity by adequate ventilation. Remove decaying leaves and fruit. Avoid overcrowding. Spray plants regularly with Benlate.

6 FOOT ROT

Foot rot is generally a disease of seedling tomatoes, but mature plants can be attacked.

Treatment: None if diseased area is large. Lift plant and burn. If plant is only slightly affected mulch stem base with moist peat and water with Cheshunt Compound; some fruit may be obtained.

Prevention: Use sterilized soil or compost for raising seedlings. Avoid overwatering. Never plant into infected soil.

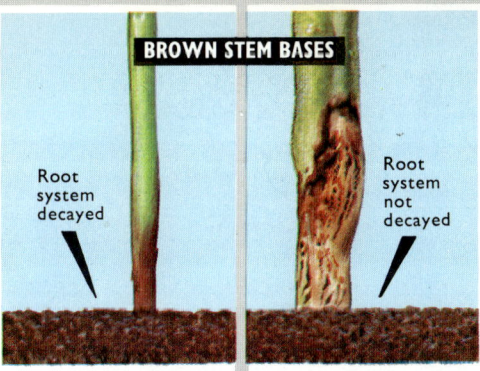

BROWN STEM BASES

Root system decayed

Root system not decayed

7 STEM ROT (Didymella)

Stem rot is a disease of mature plants. Lower leaves turn yellow and a sunken, brown canker appears at the base of the stem. Black dots develop in this cankered area. Disease may spread to other parts of the stem.

Treatment: None. Destroy badly affected plants and spray Benlate on to the stem bases of the remaining plants. If plant is only slightly affected cut out diseased area and paint with Benlate solution.

Prevention: Sterilize greenhouse and equipment between crops.

TINY MOTHS UNDER LEAVES

8 HORMONE DAMAGE

Traces of lawn weedkiller can cause severe distortion. Leaves are fern-like and twisted, stems and leaf stalks are also twisted. Similar in appearance to a virus disease, but spiral twisting is more pronounced. Fruit is plum-shaped and hollow. Avoid trouble by treating the lawn on a still day and by never using weedkiller equipment for other plants.

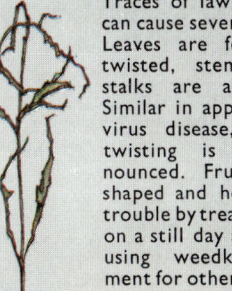

SWELLINGS ON ROOTS

Potato cyst eelworm

Root knot eelworm

9 GREENHOUSE WHITEFLY

The most widespread of all tomato pests. Both the adults and larvae suck sap from the leaves which become pale and curled. Foliage and fruit are rendered sticky; black mould grows on this honeydew, thereby disfiguring the surface.

Treatment: Not easy to control. Spray with Crop Saver or Sprayday at 3 day intervals until the infestation has been cleared. Best results are obtained by spraying in the morning or evening.

Prevention: No practical method available.

Golden Rules for staying out of trouble

Use sterilized compost when growing under glass. Don't sow or plant too early. Pay careful attention to ventilation and watering. Feed regularly and spray early against disease.

10 EELWORM

Growth is stunted and leaves are discoloured and wilted. Foliage may be purplish on the underside. Roots bear either tiny white cysts (potato cyst eelworm) or large brown swellings (root knot eelworm).

Treatment: None. Destroy infested plants.

Prevention: Do not grow tomatoes in infested soil for at least 6 years.

BROWN-STREAKED TISSUE

11 VERTICILLIUM WILT

Leaves wilt in hot weather, appearing to recover on cool evenings. Lower leaves turn yellow. If you cut open the lower stem the tell-tale signs of wilt are revealed. Brown streaks run through the stem tissue.

Treatment: Mulch around stem so new roots can form. Drench soil with Benlate solution. If possible keep at 75° F for about 2 weeks.

Prevention: Do not grow tomatoes in infected soil.

YELLOWING BETWEEN VEINS

12 MAGNESIUM DEFICIENCY

Discoloration begins on lower leaves and moves upwards until all foliage is affected. Yellow areas may turn brown. A common and serious disorder which is made worse, not better, by standard feeding.

Treatment: Spray with magnesium sulphate (see page 4) or magnesium sachet with Bio Tomato Food.

Prevention: No practical method available.

DARK BROWN BLOTCHES

13 POTATO BLIGHT

Blight can be a devastating disease of outdoor tomatoes in wet weather. The first signs are brown areas on the edges of the leaves. The patches spread until the leaves are killed. Stems show blackened patches.

Treatment: None, once the disease has firmly taken hold.

Prevention: Spray with Dithane as soon as the plants have been stopped. Repeat every 2 weeks if the weather is damp.

Tomatoes continued

FRUIT TROUBLES

14 BLOSSOM END ROT

Leathery dark-coloured patch occurs at the bottom of the fruit. It is a frequent problem where growing bags are used.

Treatment: None.

Prevention: Never let the soil or compost dry out, especially when the fruit is swelling.

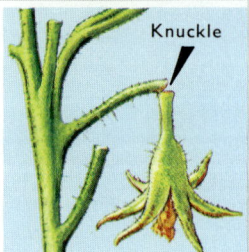

15 BLOTCHY RIPENING

Parts of the fruit remain yellow or orange and fail to ripen. The cause is usually too much heat or too little potash.

Treatment: None.

Prevention: Apply Coolglass and control heat of greenhouse. Feed with Bio Tomato Food. Water regularly.

16 BLOSSOM DROP

Flowers sometimes wither and break off at the knuckle. Pollination has not taken place, and the cause is usually dryness at the roots and in the air.

Treatment: None.

Prevention: Water regularly and spray flowers in the morning. Tap plants to aid pollination.

Knuckle

17 SUN SCALD

Pale brown, papery skinned depression occurs on the side of fruit facing the glass. Papery patches occur on leaves. Exposure to bright sunlight is the cause.

Treatment: None.

Prevention: Paint glass with Coolglass. Damp down adequately, but do not spray the plants at midday.

18 GHOST SPOT

Grey mould spores fall on or splash on to fruit. Small, transparent rings ('water spots') are formed.

Treatment: None. Affected fruit can be eaten.

Prevention: Provide good ventilation. Do not splash developing fruit when watering. Control grey mould.

19 DRY SET

Growth of the fruitlet ceases when it reaches the size of a match-head. The trouble is due to the air being too hot and dry when pollination is taking place.

Treatment: None.

Prevention: Spray the plants daily with water in the morning or evening.

20 GREENBACK

Area around the stalk remains hard, green and unripe. The cause is too much sunlight or too little potash.

Treatment: None.

Prevention: Apply Coolglass. Control heat of greenhouse. Feed regularly with Bio Tomato Food. Resistant varieties are available.

21 HOLLOW FRUIT

There are several causes of hollow fruit – poor conditions for pollination (air too hot, too cold or too dry), too little potash in the soil or damage by a hormone weedkiller.

Treatment: None.

Prevention: Avoid factors listed above.

22 TOMATO MOTH

Large green or brown caterpillars tunnel into fruit and stems. Young caterpillars eat holes in leaves.

Treatment: Too late for effective treatment at this stage. Destroy fruit.

Prevention: Spray with Crop Saver or Fenitrothion when small caterpillars and holes appear on leaves.

23 POTATO BLIGHT

Brown, shrunken area appears on fruit. The affected tomato is soon completely rotten. Infection may develop during storage.

Treatment: None, Destroy fruit.

Prevention: Protect fruit by spraying against potato blight as soon as it appears on the leaves (see page 29).

24 SPLIT FRUIT

A common complaint, both outdoors and under glass. It is caused by heavy watering or rain after the soil has become dry around the roots. This sudden increase in size causes the skin to split.

Treatment: None.

Prevention: Keep the roots evenly moist.

25 BUCKEYE ROT

Brown concentric rings around a grey spot on unripe fruit. Spores splash up from soil on to bottom trusses.

Treatment: None. Remove and destroy infected fruits.

Prevention: Tie up lower trusses to prevent splashing. Apply a peat mulch. Water carefully.

Turnips, Swedes & Radishes

WHAT'S WRONG...

	Symptom	Likely Causes
Seedlings	— eaten	**Birds** or **Slugs** (see page 11) or **Flea beetle** or **Cutworm** (see page 12)
	— toppled over	**Damping off** (see page 5)
	— peppered with small holes	**Flea beetle** (see page 12)
	— severed at ground level	**Cutworm** (see page 12)
Leaves	— swollen, distorted ('Crumple leaf')	**Swede midge** (see page 12)
	— white floury coating	**Powdery mildew** (see page 7)
	— greyish mould on underside	**Downy mildew** (see page 27)
	— white spots	**White blister** (see page 11)
	— yellowing; black veins	**4**
	— dark green, raised spots	**1**
	— infested with greenfly	**Mealy aphid** (see page 12)
	— holed	**Cabbage caterpillar** (see page 11) or **Slugs** (see page 11) or **Flea beetle** (see page 12) or **Diamond-back moth** (see page 13)
Roots	— tunnelled, maggots present	**Cabbage root fly** (see page 10)
	— swollen outgrowths	**Club root** or **Gall weevil** (see page 10)
	— covered with purple mould	**Violet root rot** (see page 15)
	— scabby patches	**Common scab** (see page 26)
	— side shoots around crown ('Many neck')	**Swede midge** (see page 12)
	— split	**Splitting** (see page 15)
	— bitter, stringy	**3**
	— woody	**Short of water or fertilizer** or **Delayed harvesting**
	— inner black ring	**4**
	— wet rot starting at crown	**2**
	— brown markings in flesh	**3**

Radishes are an easy crop to grow and only flea beetle is likely to trouble you. Both turnips and swedes can be affected by a host of problems but only four are common — club root, flea beetle, powdery mildew and soft rot.

DARK GREEN SPOTS

1 TURNIP MOSAIC VIRUS

An infectious and damaging disease of turnips, which is fortunately uncommon. Young leaves are twisted and mottled; it may be fatal to young plants. Tell-tale sign is the presence of dark green, raised spots on the leaves.

Treatment: Destroy affected plants, as this disease can lead to soft rot.

Prevention: Spray with Topgard Systemic Liquid or Crop Saver to control greenfly.

2 SOFT ROT

A wet and slimy rot, beginning at the crown, can occur in both the growing crop and in stored roots. The outer skin of the roots remains firm. A tell-tale sign is the collapse of the foliage. Soft rot can be serious, especially in a wet season, and it is essential to remove affected plants immediately. To avoid trouble next season make sure the soil is well drained, avoid over-manuring, be careful not to injure roots when hoeing and never store damaged turnips or swedes. Practise crop rotation.

Golden Rules for staying out of trouble
Leave several months between digging and planting. Rake in Bromophos. Quick growth is essential, so water and feed. Spray at the first sign of flea beetle. Store roots in a dry, frostproof place.

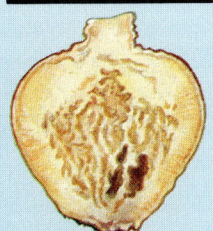

BROWN MARKINGS

3 BROWN HEART

Greyish-brown rings run through the flesh. Affected areas become water-soaked. This disease is much more likely to attack swedes than turnips, and is usually restricted to light soils in a dry season. Affected roots are bitter. The cause is boron deficiency.

Treatment: None.

Prevention: If soil is known to be boron deficient, apply 1 oz borax per 20 sq. yards before planting – take care not to overdose.

OUTER BLACK RING

4 BLACK ROT

Above ground the symptoms of black rot are yellow leaves with black veins (see page 11). If an affected root is cut across a tell-tale ring of black dots can be seen just below the skin. Attacks are worst in a warm, wet summer on poorly drained soil.

Treatment: None. Lift diseased plants and burn.

Prevention: Practise crop rotation. Make sure the soil is well drained.

Other Vegetables

Asparagus

ASPARAGUS BEETLE
Grubs and adult beetles attack both stems and foliage. Stems are eaten; leaves are stripped. Beetle is easily recognized by square orange markings on $\frac{1}{4}$ in. long black body. Spray with Liquid Derris or Malathion at first sign of attack.

SLUGS
Spears are gnawed, making them unfit for table use. See page 5.

VIOLET ROOT ROT
Most serious asparagus disease. Roots covered with purplish mould; leaves turn yellow and die. If attack is severe make a new bed on a fresh site. Do not grow root vegetables on the affected area for at least 3 years.

FROST
Young shoots may turn black and die if a severe frost occurs in late spring. Destroy affected shoots. Cover bed with sacking if a hard frost is expected.

RUST
Reddish-brown spots appear on the leaves during the summer. Spraying is not effective so cut down and burn affected shoots as soon as the first spots are seen.

WIND ROCK
Roots are loosened if stems are left unsupported on an exposed site. This can lead to rotting, so some summer support should be provided if windbreaks are absent.

SPINDLY SPEARS
Thin shoots, about $\frac{1}{8}$ in. across, are sometimes produced instead of typical thick spears. The most likely cause is prolonged cutting in the previous season. Spears should not be harvested after mid June. Other possible causes are cutting too soon after planting and failing to feed the bed during the summer.

Aubergine

RED SPIDER MITE
Pale mottling occurs on the upper surface of leaves. Tiny mites can be found on the underside. Spray thoroughly with Derris or Malathion at the first sign of attack. Never spray in bright sunlight nor when the plants are dry at the root.

Capsicum

RED SPIDER MITE
Sweet peppers are extremely susceptible to red spider mite, and careful watch should be kept for pale mottling and bronzing of the leaves. Tiny mites can be found on the underside. Spray thoroughly with Derris or Malathion at the first sign of attack. Never spray in bright sunlight nor when the plants are dry at the root.

Chicory

SOIL PESTS
Cutworms, wireworms and swift mot[h] caterpillars can be troublesome, so rak[e] Bromophos into the soil before sowin[g] seed.

BITTER HEADS
Forced heads ('chicons') are sometime[s] yellow and bitter. The cause is failing t[o] keep them in absolute darkness durin[g] the forcing period.

Globe Artichoke

ROOT APHID
Brownish 'greenfly' attack the root[s] which become covered with a wax[y] secretion. Growth is stunted. If roo[t] aphid attack is identified, wate[r] around the plants with spray-strengt[h] Malathion.

PETAL BLIGHT
A serious but uncommon disease affecting the young heads. Brow[n] spots rapidly join together, so tha[t] the heads turn completely brown. Remove and burn affecte[d] artichokes. If petal blight is known to be a nuisance on the sit[e] spray with Dithane when buds first appear. Repeat at for[t] nightly intervals.

APHID
Both blackfly and greenfly attack developing flower head[s] Spray with Crop Saver as soon as the first attacks are see[n]

Jerusalem Artichoke

DAMAGED TUBERS
Hollowed-out tubers indicate attac[k] by either swift moth caterpillar[s] chafer grubs or slugs. You can avoi[d] trouble by raking Bromophos into th[e] soil before planting tubers and b[y] sprinkling Slug Pellets around th[e] growing crop.

SCLEROTINIA ROT
The base of stems are attacked, and may show fluffy whit[e] mould. If a rotten stem is cut open, large black cyst-lik[e] bodies will be seen inside. Tubers may also be affected. Se[e] page 15 for control measures.

Mint

RUST
Rust is by far the most serious disorde[r] of mint. In spring, small orange spot[s] appear on swollen shoots. Later in th[e] season affected leaves dry up and fal[l] There is no cure. The best plan is t[o] make a new bed in another part of th[e] garden, using healthy mint plant[s] Alternatively cover diseased plants wit[h] straw in autumn and burn off the to[p] growth.

Mustard & Cress

DAMPING OFF

This is the only problem of this easily-grown crop, as it is cut at the seedling stage. Grow the seeds on sacking, blotting paper or sterilized compost and make sure the atmosphere is cool and not too moist. It is vital to grow fresh seed to avoid uneven germination which leads to disease.

Salsify

WHITE BLISTER

Shiny white blisters on the leaves; growth is stunted and root development is affected. Cut off and burn diseased leaves.

SHORT ROOTS

Long roots require deep sandy soil which contains plenty of well-rotted organic matter.

Parsley

POOR GERMINATION

It is usual for parsley seed to take 4–8 weeks to germinate so the long delay before seedling emergence is not abnormal. To speed up germination use fresh seed and line the drill with moist peat before sowing.

LEAF SPOT

Small angular spots, brown at first and then nearly white, occur on the leaves. Spray with Benlate at the first sign of disease; repeat as necessary. If left unchecked the plants will be seriously damaged in wet weather. It will then be necessary to sow fresh seed in a different part of the garden.

Seakale

BLACK ROT

Leaves turn yellow with distinctive black veins. There is no cure, and so diseased plants should be destroyed. Avoid trouble by growing seakale in well-drained soil and practise crop rotation.

CLUB ROOT

Affected roots are swollen and misshapen; leaves wilt in sunny weather. Avoid trouble by making sure the soil is adequately limed and well drained. See page 10 for further details of treatment and prevention.

Rhubarb

CROWN ROT

The terminal bud rots and the tissue below the crown decays. The sticks are spindly and dull coloured. There is no cure, so badly infected plants should be dug out and burnt. If several plants are attacked, make a new bed with healthy plants in a different part of the garden. Make sure the soil is well-drained.

HONEY FUNGUS

Tell-tale sign is the presence of white streaks in the brown dead tissue of the crown. Orange toadstools appear around the affected plants. This is an extremely serious disease of trees, shrubs and perennial vegetables, so it is vital to dig out and burn diseased roots.

BLACK BEAN APHID

Colonies of blackfly appear on the leaves; spray with Topgard Systemic Liquid or Crop Saver as soon as the first attacks are seen.

LEAF SPOT

Small spots appear on the leaves. Central area of each spot may fall out, producing a 'shot hole' effect. Neither yield nor quality is affected so control measures are unnecessary.

Sweet Corn

SMUT

Large galls ('smut balls') appear on the ears and stalks in hot, dry weather. These galls should be cut off and burnt before they burst open, releasing a mass of black spores. Burn all plants after harvesting and do not grow sweet corn on the site for at least 3 years.

POORLY FILLED COBS

Cobs containing only a few kernels are usually a sign of poor pollination. To prevent this complaint always plant sweet corn in blocks rather than in single rows.

FRIT FLY

Frit fly maggots bore into the growing points of corn seedlings, which develop twisted and ragged leaves. Growth is stunted and undersized cobs are produced. Control measures are not generally worth while but seedlings can be protected by dusting them with Lindane.

TOUGH KERNELS

Cooked corn should be tender and sweet; tough starchy kernels can be due to three separate causes. Waiting too long before harvesting is the most likely reason, but both keeping the cobs in store too long and boiling the cobs for more than 5 minutes can result in toughness.

Vegetable Doctor's Dictionary

ACARICIDE
A chemical used to control mites, such as red spider mite.

BLANCHING
Excluding light from stems of celery, leek etc. to make them more palatable.

BLEEDING
The loss of sap from plant tissues after having been cut.

BLIND
The loss of the growing point, which results in stoppage of growth. Watch for this condition in brassica seedlings.

BOLTING
Premature running to seed. Often caused by a check to growth arising from drought or starvation.

CANKER
A vague term for a localized area of sunken dry rot on a plant stem. Also used to describe a serious parsnip disease.

CATERPILLAR
The larva of a moth or butterfly.

CHLOROSIS
Loss of green colour from the foliage, which turns yellow or white.

CURD
The tight mass of young flower buds which make up the heads of cauliflower and broccoli.

DAMPING DOWN
Watering the floor and benches of a greenhouse, usually in warm weather, to create a humid atmosphere.

DAMPING OFF
Decay of young seedlings at ground level.

DRAWN
Excessively tall, weak and thin growth, caused by plants being grown in shade or too closely together.

DROUGHT
A prolonged period of dry weather. Officially it is any period of 14 days or more without measurable rainfall.

EARTHING UP
Drawing up soil around the stems of plants.

FOLIAR FEED
A solution of plant nutrients which is rapidly absorbed into the sap stream after spraying on to the leaves.

FOOT ROT
Decay of the tissues at the base of the stem.

FORCING
The hastening of growth by providing extra warmth. Some vegetables, such as chicory, are forced in complete darkness.

FRIABLE
Earth which is crumbly and easily worked.

FUNGICIDE
A chemical used to control plant diseases caused by fungi.

GALL
An abnormal outgrowth which appears on plants, caused by the action of insects or bacteria.

GROWING BAG
Plastic bag filled with compost in which vegetables are grown.

HAULM
Another name for the shoots of some vegetables (peas, beans, potatoes etc.).

HEART ROT
A general term for various internal decays of root crops and celery.

HONEYDEW
Sticky, sugary secretion deposited on plants by insects such as aphids and whitefly.

HOST PLANT
Species of plant on which a particular insect or fungus can live.

IMMUNE
Resistant to a specific disease. Note that immune varieties are not resistant to all diseases.

INORGANIC
The scientific meaning is a pesticide or fertilizer which does not contain carbon. The popular meaning is a chemical which is not obtained from a source which is – or has been alive.

INSECTICIDE
A chemical used to control insect pests.

LARVA
The first and active feeding stage of an insect, which hatches from the egg. Caterpillars later become moths or butterflies; maggots become flies.

LEAF SCORCH
The browning and the withering of the leaf edge, due to shortage of potash and exposure to strong sunlight under glass.

LESION
A localised depression or hollow which is a symptom of disease.

MILDEW
A general term for a large number of unrelated diseases which produce a white or pale-coloured mould on the leaves or stems.

MITE
Tiny 8 legged insect-like pest.

MOSAIC
A form of virus disease which causes pale green or yellow mottling of the leaves.

MOTTLE
Spots or blotches of pale green or yellow which are scattered over the leaf surface.

MULCH
A surface layer of organic matter, used to suppress weeds and conserve moisture.

MYCELIUM
White threads produced by a fungus.

NEMATODE
The scientific name for an eelworm.

ORGANIC
A substance obtained from a source which is or has been alive.

PARASITE
An organism which lives and feeds on another organism.

PREDATOR
An insect which preys and feeds on another.

RODENTICIDE
A chemical used to control rats and mice.

RUST
Leaf swellings or spots which contain masses of yellow, red or brown spores.

SCAB
Disease which causes surface swellings which are rough and scurfy.

SCURFY
A surface condition where a scaly or crusty layer is formed.

SECONDARY ATTACK
An infection or insect attack which occurs on tissue already damaged by another trouble. Millepedes, woodlice and grey mould frequently attack plants in this way.

SET
The fertilization of the flower — the start of fruit development.

SETS
Tubers (potatoes) or bulbs (onions and shallots) used for planting.

SMUT
Disease producing swellings which contain masses of black spores.

STERILIZED SOIL
A rather misleading term, as steam – or chemically-sterilized soil is only partially sterilized. Harmful organisms have been killed but helpful bacteria have been spared.

STOPPING
The removal of the growing point. This is usually done by pinching out the tip of the plant between finger and thumb.

SYSTEMIC
A fungicide, insecticide or fertilizer which is able to penetrate the leaves and enter the sap stream.

TRUSS
A cluster of fruit at the end of a stem, as in tomatoes.

TUBER
The edible swelling of an underground stem, such as potato and Jerusalem artichoke.

VECTOR
The carrier of a disease.

VENTILATION
The creation of air movement by the opening of the ventilators of a greenhouse so as to reduce the temperature and humidity.

VIRUSES
Invisible organisms smaller than bacteria. They are carried in the sap and may be transmitted from plant to plant by tools, hands or insects.

WEEVIL
A beetle with a distinct 'snout'.

WINDBREAKS
A natural or artificial structure which diverts or reduces the force of the wind.

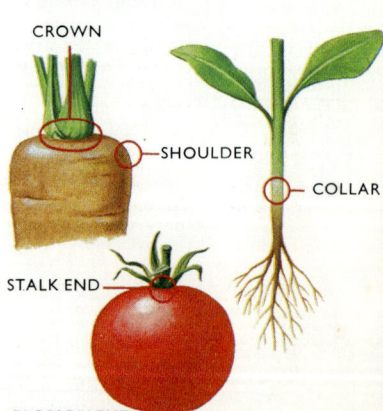

CROWN · SHOULDER · COLLAR · STALK END · BLOSSOM END